PARTIAL DIFFERENTIAL EQUATIONS

Series on Soviet and East European Mathematics – Vol. 2

PARTIAL DIFFERENTIAL EQUATIONS

A V Bitsadze

Published by

World Scientific Publishing Co. Pte. Ltd.

P O Box 128, Farrer Road, Singapore 9128

USA office: Suite 1B, 1060 Main Street, River Edge, NJ 07661

UK office: 73 Lynton Mead, Totteridge, London N20 8DH

PARTIAL DIFFERENTIAL EQUATIONS

ISBN 981-02-0593-7

Printed in Singapore by JBW Printers & Binders Pte. Ltd.

Preface

This book is based on the lectures on partial differential equations given by the author to students of the fourth year in Department of Computational Mathematics and Cybernetics of Moscow M. V. Lomonosov State University for several sessions in recent years.

The theoretical foundations of partial differential equations are explained rigorously and clearly in such a way that their importance in applications is also taken into account. The materials have been arranged to promote the students' interest to mathematical experiments where partial differential equations are considered importance.

A. Bitsadze

Contents

Chapter I. Modelling Some Physical Phenomena in Terms of Partial Differential Equation

§1.1 The Concept of Partial Differential Equations

1.1.1. Suppose that D is a domain in an n-dimensional Euclidean space E_n of points x with coordinates $x_1, x_2 \ldots, x_n$, where $n \geq 2$, and $F(x, \ldots, p_{i_1 \ldots i_n}, \ldots)$ is a real function given for all $x \in D$ and all possible real variables $p_{i_1 i_2 \ldots i_n}$ with non-negative indices $i_1, \ldots, i_n$ for which $\sum_{j=1}^{n} i_j = k, k = 0, \ldots, m, m \geq 1$.

Under the assumption that at least one of all the partial derivatives $\frac{\partial F}{\partial p_{i_1 \ldots i_n}}$ is different from zero for $k = m$, an equality of the form

$$F\left(x, \ldots, \frac{\partial^k u}{\partial x_1^{i_1} \ldots \partial x_n^{i_n}}, \ldots\right) = 0 \tag{1}$$

is called a *partial differential equation* (or simply *an equation with partial derivative*) of order m with respect to some unknown function $u(x) = u(x_1, \ldots, x_n)$ and the left part of this equality is called a *partial differential operator* of order m.

The real function $u(x)$, which is continuous together with all its partial derivatives, entering this equation and converting it into an identity, is called a *regular solution* of this equation.

1.1.2. Equation (1) is said to be *linear* if F is dependent upon all $p_{i_1 \ldots i_n}$ linearly. If the function F depends on $p_{i_1 \ldots i_n}$ linearly only for $k = m$, then Eq. (1) is called *quasilinear*.

A linear partial differential equation of order m can always be written in the form

$$\sum_{k=0}^{m} \sum_{i_1 \ldots i_n} A^k_{i_1 \ldots i_n}(x) \frac{\partial^k u}{\partial x_i^{i_1} \ldots \partial x_n^{i_n}} = f(x) \,,$$

$$k = \sum_{j=1}^{n} i_j \,, \quad i_j \geq 0 \,, \tag{2}$$

where $A^k_{i_1 \ldots i_n}(x)$ and $f(x)$ are functions given in the domain D. For linear second order equation it is more convenient to write, instead of (2),

$$\sum_{i,j=1}^{n} A_{ij}(x) \frac{\partial^2 u}{\partial x_i \partial x_j} + \sum_{i=1}^{n} B_i(x) \frac{\partial u}{\partial x_i} + C(x) u = f(x) \tag{3}$$

The linear equations (2) and (3) are called *homogeneous* or *nonhomogeneous* depending on whether or not the function $f(x)$ is identically zero on the whole D.

Equation (3) stops being a second order equation at points $x \in D$ where all the coefficients A_{ij} are equal to zero. That is, the order of Eq. (3) degenerates at these points.

1.1.3. If F is an N-dimensional vector $F = (F_1, \dots, F_N)$ with components $F_i(x, \dots, p_{i_1 \dots i_n}, \dots), i = 1, 2, \dots, N$, which depends on $x \in D$ and the N_1-dimensional vectors

$$p_{i_1 \dots i_n} = \left(p^1_{i_1 \dots i_n}, \dots, p^{N_1}_{i_1 \dots i_n}\right),$$

the vector equality (1), where $u = (u_1, \dots, u_{N_1})$ is an N_1-dimensional vector, is called a *system of partial differential equations* with respect to the unknown functions $u_1, \dots, u_{N_1}$ or the unknown vector $u = (u_1, \dots u_{N_1})$.

For the union of non-negative integer indices $i_1, i_2, \dots, i_n$ of variables $p^l_{i_1 \dots i_n}, l = 1, 2, \dots, N_1$, we accept the notation $\sum_{j=1}^{n} i_j = k_l, \tilde{k}_l = 0, \dots, m_l, m_l \geq 1$. The largest order of the derivatives of the unknown functions entering into the given equation of the system is called the *order of the equation.*

In the system of equations it is not necessary at all to require the number N of the equation to be equal to the number N_1 of the unknown functions, or the orders of all equations in the given system to be the same.

If the variables that enter the left-hand side of Eq. (1) are complex and, for $x_k = y_k + iz_k$, by $\frac{\partial}{\partial x_k}$ we mean $\frac{1}{2}(\frac{\partial}{\partial y_k} - i\frac{\partial}{\partial z_k})$, then evidently Eq. (1) is equivalent to a system of equations.

§1.2 Partial Differential Equations Used as Euler-Lagrange Equations in Variational Problems

1.2.1. Certain phenomena that involve flows in mixed media and fields, or that which are investigated in physics, natural sciences and technology, in general, have been modelled mathematically with partial differential equations, which are actually differential descriptions of the conservation laws inherent in these phenomena.

One of the methods to obtain partial differential equations that comes from the conservation laws originates from the variational calculus.

We will assume that (1) the investigated phenomenon has been described by real functions $u_1(x), \ldots, u_m(x)$ with $x = (x_1, \ldots, x_n) \in G$, which are sufficiently smooth in some domain G of space E_n and continuous on $G \cup \partial G$, having preassigned values on the $(n-1)$-dimensional boundary $S = \partial G$ of the domain G; (2) the variable whose conservation is under consideration is the functional

$$E(u) = \int_G \phi(x_1, \ldots, x_n, u_1, \ldots, u_m, p_{11}, \ldots, p_{1n}, \ldots, p_{m1}, \ldots, p_{mn}) dx ,$$

where ϕ is a given real, and sufficiently smooth function of its arguments, dx is the volume element of G and $p_{ki} = \frac{\partial u_k}{\partial x_i}$; (3) the *conservation law* which is inherent in an investigated phenomenon is contained in the requirement that the system of functions $u_1(x), \ldots, u_m(x)$ give the functional $E(v)$ a stationary (extreme) value among all admissible systems of functions, i.e., among all possible systems of functions $v_1(x), \ldots, v_m(x)$, where $x \in G$, which are close to $u_1(x), \ldots, u_m,(x)$ in the definite sense that for them $E(v)$ exists and they have the same values as $u_1, \ldots, u_m$ on S (*Hamilton's principle*).

Without loss of generality, assuming that the system of functions admissible for the functional $E(v)$ can be presented in the form $v_k(x) = u_k(x) + \varepsilon_k h_k(x), k = 1, \ldots, m$, where ε_k are arbitrary real constants and $h_k(x)$ are arbitrary admissible functions which convert into zero on S, we arrive at the conclusion that the variable $I(\varepsilon_1, \ldots, \varepsilon_m) = E(u_1 + \varepsilon_1 h_1, \ldots, u_m + \varepsilon_m h_m)$, as a function of $\varepsilon_1, \ldots, \varepsilon_m$, achieves its own stationary (extremal) value when $\varepsilon_1 = \ldots = \varepsilon_m = 0$, i.e.,

$$\left(\frac{\partial I}{\partial \varepsilon_k}\right)_{\substack{\varepsilon_1=0 \\ \vdots \\ \varepsilon_m=0}} = \int_G \left(\frac{\partial \phi}{\partial u_k} h_k + \sum_{i=1}^{n} \frac{\partial \phi}{\partial p_{ki}} \frac{\partial h_k}{\partial x_i}\right) dx = 0 , k = 1, \ldots, m .$$

Hence, taking the obvious identity

$$\sum_{i=1}^{n} \frac{\partial \phi}{\partial p_{ki}} \frac{\partial h_k}{\partial x_i} = \sum_{i=1}^{n} \frac{\partial}{\partial x_i} \left(\frac{\partial \phi}{\partial p_{ki}} h_k\right) - h_k \sum_{i=1}^{n} \frac{\partial}{\partial x_i} \frac{\partial \phi}{\partial p_{ki}}$$

into account, and using the *Gauss-Ostrogradskii formula*

$$\int_G div\ w(x) dx = \int_s w \cdot \nu\ ds , \tag{G-O}$$

which is valid for any vector $w = (w_1, \ldots, w_n)$ which is smooth in domain G under the requirement of the boundedness of G and the smoothness of its boundary S, we obtain that

$$\int_G \left(\frac{\partial \phi}{\partial u_k} - \sum_{i=1}^{n} \frac{\partial}{\partial x_1} \frac{\partial \phi}{\partial p_{ki}} \right) h_k dx = 0 \ , k = 1, \ldots, m \ .$$

In turn, because of the arbitrariness of functions $h_k(x), k = 1, \ldots, m$, it follows from these equalities that the system of functions $u_1(x), \ldots, u_m(x)$ should be the solution of the system of partial differential equations

$$\frac{\partial \phi}{\partial u_k} - \sum_{i=1}^{n} \frac{\partial}{\partial x_i} \frac{\partial \phi}{\partial p_{ki}} = 0 \ , \quad k = 1, \ldots, m \ ,$$

known as *Euler-Lagrange equations* for the functional $E(u)$.

1.2.2. The scheme of deriving Euler-Lagrange equations as indicated above is also suitable for cases where the independent variables $x_1, \ldots, \ldots, x_n$ and unknown functions $u_1, \ldots, u_m$ are regarded as points in sufficiently smooth n and m-dimensional manifolds.

Let M and M^* be smooth manifolds with local coordinates $(x^1, \ldots, x^n)$, $(y^1, \ldots, y^m)$ and metric quadratic forms $ds^2 = \sum_{i,j=1}^{n} g_{ij} dx^i dx^j$, $ds_*^2 = \sum_{\alpha,\beta=1}^{m} g^*_{\alpha\beta} dy^\alpha dy^\beta, g_{ij} = g_{ji}, g^*_{\alpha\beta} = g^*_{\beta\alpha}$, respectively, while u be a smooth mapping of M onto M^*. We denote by g^{ij} the ratio of the cofactor of the element g_{ij} in matrix $||g^{ij}||$ to $\det ||g_{ij}|| = g$.

The expression

$$\mathcal{L}(u) = \frac{1}{2} \sum_{i,j=1}^{n} \sum_{\alpha,\beta=1}^{m} g^*_{\alpha\beta} \frac{\partial u_\alpha}{\partial x^i} \frac{\partial u_\beta}{\partial x^j} g^{ij}$$

is usually called *Lagrange's energy density.* When M is assumed to be compact, the functional

$$E(u) = \int_M \mathcal{L}(u) dm \ ,$$

where dm is a volume element of M, is called an *energy integral.*

Euler-Lagrange equations of $E(u)$ can be expressed in the form

$$\Delta u_\alpha + \sum_{\beta,\gamma=1}^{m} \sum_{i,j=1}^{n} \Gamma^{*\alpha}_{\beta\gamma} \frac{\partial u_\beta}{\partial x^i} \frac{\partial u_\gamma}{\partial x^j} g^{ij} = 0, \alpha = 1, \ldots, m, \tag{4}$$

where

$$\Delta = -\frac{1}{\sqrt{g}} \sum_{i,j=1}^{n} \frac{\partial}{\partial x^i} \left(\sqrt{g} g^{ij} \frac{\partial}{\partial x^i} \right)$$

is the *Beltrami differential operator*, and $\Gamma^{*\alpha}_{\beta\gamma}$ is Christoffel's symbol on M^*.

If $g_{ij} = 0, i \neq j, g^{ii} = -1$, then the Beltrami operator coincides with the *Laplace operator* which obviously can also be written in the terms of $\Delta = divgrad$ in the orthogonal Cartesian coordinates.

Now we turn our attention to some examples of modelling investigated phenomena with the help of partial differential equations.

§1.3 Oscillations of a Membrane

1.3.1. An elastic two-dimensional solid medium (material surface) which has the shape of planar domain G in the rest state and whose potential energy in process of oscillation is proportional to the increment of its area is called a *membrane*. The elasticity of a membrane is expressed through the requirement that it accomplishes the oscillating motion in a finite period of time after the external action which leads it out of the rest state has stopped.

1.3.2. As a system of reference, we shall accept the orthogonal Cartesian coordinates x, y, t and denote the corresponding Euclidean space by E_3. We suppose that the domain G lies in the plane of variables x, y and that the *deflection of the membrane* $u(x, y, t)$, namely the vertical displacement of a point $(x, y) \in G$, is a sufficiently smooth function. We shall assume that the membrane is fixed along its edges and its oscillation is small in the sense that we can neglect all the over 2 powers of u, u_x, u_y in computations

Since the area σ of a membrane which is brought out from the rest state at instant t is given by the formula

$$\sigma = \int_G \sqrt{1 + u_x^2 + u_y^2}\, dx dy \approx \int_G \left(1 + \frac{1}{2} u_x^2 + \frac{1}{2} u_y^2 \right) dx dy \ ,$$

and its area in rest state is

$$|G| = \int_G dxdy \ ,$$

the potential energy $E_{(p)}$ has expression

$$E_{(p)} = \frac{1}{2}\mu \int_G (u_x^2 + u_y^2) dxdy \ ,$$

where the proportional coefficient μ has the name of *tension of membrane*.

The *kinetic energy of an oscillating membrane* is given by the formula

$$E_{(k)} = \frac{1}{2} \int_G \rho u_t^2 \, dxdy \ ,$$

where ρ is the surface mass density of the membrane, and u_t is its rate of displacement.

By virtue of Hamilton's principle, the oscillation of a membrane takes place in such a way that the function $u(x, y, t)$ gives a stationary value to the integral

$$\int_{t_1}^{t_2} (E_{(k)} - E_{(p)}) dt = \frac{1}{2} \int_{t_1}^{t_2} dt \int_G \left[\rho u_t^2 - \mu(u_x^2 + u_y^2)\right] dxdy \ , \qquad (5)$$

where (t_1, t_2) is the interval of time for the oscillation.

In the considered example,

$$\mathcal{L}(u) = \frac{1}{2}\left[\rho u_t^2 - \mu(u_x^2 + u_y^2)\right]$$

plays the role of a Lagrangian, the role of energy integral is played by the expression (5), and the Euler-Lagrange equation (4) for the functional (5) is now written in the form

$$\frac{\partial}{\partial t}(\rho u_t) - \frac{\partial}{\partial x}(\mu u_x) - \frac{\partial}{\partial y}(\mu u_y) = 0 \ . \qquad (6)$$

The equality (6) concerning $u(x, y, t)$ is a linear partial differential equation of the second order. If ρ and μ are constant, we obtain from (6)

$$\frac{1}{a^2}\frac{\partial^2 u}{\partial t^2} - \frac{\partial^2 u}{\partial x^2} - \frac{\partial^2 u}{\partial y^2} = 0 \ , \qquad (7)$$

where $a^2 = \mu/\rho$. The constant a has the name of *rate of sound propagation.* Without loss of generality we can suppose that it is equal to unity.

1.3.3. Assuming that u is independent of t, i.e., in the bending state the membrane is in equilibrium which is described by the equation $u = u(x, y)$, we obtain from (7) the *Laplace equation*

$$\Delta u = 0 \,, \quad \text{with} \quad \Delta = \frac{\partial^2}{\partial x^2} + \frac{\partial^2}{\partial y^2} \,,$$

being the Euler-Lagrange equation of the functional

$$D(u) = \int_g (u_x^2 + u_y^2) dx dy \,,$$

which is called the *Dirichlet integral.*

Under the corresponding selection of scale, the Dirichlet integral expresses the potential energy of a membrane in equilibrium and with deflection $u(x, y)$.

§1.4 Propagation of Heat

1.4.1. The process of *heat propagation* in a medium with mass density ρ, under the condition of a given *specific heat* C and a *heat-conduction coefficient* k, can be modelled mathematically as follows. Let $u(x, t)$ be the temperature of the medium at point x and time t, and D be an arbitrary region of the medium containing the point x. It is assumed that no source or sink of heat exists in D. By S we denote the boundary of D.

1.4.2. If ds and ν are the area element of S and the outward normal on S, then, under the condition that $u(x, t)$ is a sufficiently smooth function, the quantity of heat, Q, which enters D through S during the interval of time (t_1, t_2) is given by the formula

$$Q = \int_{t_1}^{t_2} dt \int_s k \frac{\partial u}{\partial \nu} ds \,,$$

by virtue of *Fourier's law.*

As a result of the inflow of the heat Q, the increment of temperature is equal to $u(x, t + dt) - u(x, t) = u_t dt$, so that

$$Q = \int_{t_1}^{t_2} dt \int_D c\rho \; u_t \; d\tau \,,$$

where $d\tau$ is the volume element.

Therefore,

$$\int_{t_1}^{t_2} dt \int_s k\frac{\partial u}{\partial \nu} ds = \int_{t_1}^{t_2} dt \int_D c\rho \, u_t \, d\tau \; . \tag{8}$$

By virtue of the Gauss-Ostrogradskii formula (G-O) we have

$$\int_s k\frac{\partial u}{\partial \nu} ds = \int_s k \; grad \; u \cdot \nu ds = \int_D div(k \; grad \; u) d\tau \; .$$

Based on this identity we rewrite the equality (8) in the form

$$\int_{t_1}^{t_2} dt \int_D [c\rho \; u_t - div(k \; grad \; u)] d\tau = 0 \; , \tag{9}$$

where the vector differential operator *grad* and the scalar differential operator *div* are taken with respect to the space variables x_1, x_2, x_3.

In view of the fact that the interval (t_1, t_2) of time and the domain D are arbitrary, we obtain from (9) that

$$c\rho\frac{\partial u}{\partial t} - div(k \; grad \; u) = 0 \; .$$

From here, assuming that ρ, c and k are constants, we obtain

$$\frac{1}{a^2}\frac{\partial u}{\partial t} - \Delta u = 0 \; , \tag{10}$$

where $\Delta = divgrad = \Sigma_{i=1}^{3} \frac{\partial^2}{\partial x_i^2}$ is the Laplacian and $a^2 = k/c\rho$.

Evidently, without loss of generality, we can now suppose that $a = 1$ and hence write Eq. (10) in the form

$$u_t - \Delta u = 0 \; . \tag{11}$$

Equation (11) is called the *heat-conduction equation.*

When $u(x, t)$ is independent of t (i.e., when the process of heat propagation is stationary), Eq. (11) reduces to Laplace equation $\Delta u = 0$.

§1.5 Equation of Linear Theory of Elasticity

1.5.1. In studying a continuous medium, one should distinguish two kinds of forces that act on it; there are the volume forces and the surface forces. Let us single out conceptually an arbitrary volume τ with boundary σ in the medium.

Volume forces act on every point of τ (for example, gravity), and surface forces act on every point of σ (for example, stress). The latter forces are the kind of force that interacts between the parts of the medium that lie in the in interior and in the exterior of σ. If there exist surface forces, whether the medium is staying in the rest state or in motion it is said to be situated in a stress state. Therefore, this kind of force is called *stress*.

A volume force ϕ applied on a point $P(x, y, z) \in \tau$, is a vector, and it is known if its three coordinates X, Y, Z (the projection on the coordinate axes, in our case of orthogonal Cartesian system) are known.

The stress ψ at every point $P(x, y, z) \in \sigma$ depends not only on the point P, but also on the orientation of the area element of the surface σ which is assigned to this point. It will be considered as known if we know (a) the orientations of three surface elements $d\sigma_{\nu_1}, d\sigma_{\nu_2}, d\sigma_{\nu_3}$ at point P where their normals ν_1, ν_2, ν_3 are not coplanar and (b) the stresses that act on each of these elements. We take the unit vectors of the coordinate axes x, y, z with surface elements $d\sigma_x, d\sigma_y, d\sigma_z$ as the ν_1, ν_2, ν_3. Each of the stresses that acts on $d\sigma_x, d\sigma_y$ and $d\sigma_z$ has three components which are simply denoted by (X_x, Y_x, Z_x), $(X_y, Y_y, X_y), (X_z, Y_z, Z_z)$, respectively. The values of these nine variables determine completely the stress ψ that acts on the point $P \in \sigma$. It can be proved that they make up the components of a second rank tensor, which is called a *tensor of stress*.

The stress ψ acting on point $P(x, y, z) \in \sigma$ is a vector with components

$$
\begin{aligned}
X_\nu &= X_x \cos \widehat{\nu x} + X_y \cos \widehat{\nu y} + X_z \cos \widehat{\nu z} \ , \\
Y_\nu &= Y_x \cos \widehat{\nu x} + Y_y \cos \widehat{\nu y} + Y_z \cos \widehat{\nu z} \ , \\
Z_\nu &= Z_x \cos \widehat{\nu x} + Z_y \cos \widehat{\nu y} + Z_z \cos \widehat{\nu z} \ ,
\end{aligned}
$$

where ν is the unit vector of the outward normal to σ at the point P.

1.5.2. In a medium the stress state arises as the result of external influence which changes the distances among points, i.e., when the medium is subjected to deformation. If a medium shows resistance to the external influence such that in the process of the influence it is deformed but after gradually stopping this influence it returns to its initial state, then it is called a *rigid elastic body*. As a result of stopping instantly the external influence, a rigid body will produce oscillatory motion for a certain interval of time.

If a point P of an elastic body D is present in a position Q after deformation, then the vector $PQ = u = (u_1, u_2, u_3)$ is called a *displacement vector* or a *shift vector*, and the scalar variables u_1, u_2, u_3 are called *displacement components* or *shift components*.

The variables $e_{xx}, e_{xy}, e_{xz}, e_{yx}, e_{yy}, e_{yz}, e_{zx}, e_{zy}, e_{zz}$ defined by the formulae

$$\begin{aligned} e_{xx} &= \frac{\partial u_1}{\partial x}, \quad e_{yx} = e_{yx} = \frac{\partial u_2}{\partial x} + \frac{\partial u_1}{\partial y}, \\ e_{yy} &= \frac{\partial u_2}{\partial y}, \quad e_{yz} = e_{zy} = \frac{\partial u_3}{\partial y} + \frac{\partial u_2}{\partial z}, \\ e_{xz} &= e_{xz} = \frac{\partial u_3}{\partial x} + \frac{\partial u_1}{\partial z}, \quad e_{zz} = \frac{\partial u_3}{\partial z}, \end{aligned} \tag{12}$$

are called *deformation components*.

In order to be able to determine the functions u_1, u_2, u_3 from (12), these deformation components should be connected among themselves by six equalities, which are called *Saint Venant's conditions of consistency*.

An elastic body is said to be *isotropic* if its properties are the same in all directions. We shall only consider isotropic elastic bodies below.

1.5.3. Assuming that the deformations are small, i.e., disregarding powers of the variables u, u_x, u_y, u_z beyond the second, and ignoring any rigid shift of the elastic body D as a unit, in the linear theory of elasticity with certain acceptable accuracy, the expression

$$E_d = G \int_D \left[\frac{m-1}{m-2} (div\ u)^2 + \frac{1}{2} (rot\ u)^2 \right] dxdydz \tag{13}$$

is taken as the energy of deformation, where the differential operators *div* and *rot* are taken with respect to the spatial variables x, y, z, G and m are completely given, constant and characteristic for the given elastic body, and, furthermore, $G > 0, m > 2$.

We denote by E_k and A respectively the kinetic energy of the elastic body D and the work carried out by volume force ϕ:

$$E_k = \frac{1}{2} \int_D \rho (u_t)^2 dxdydz , \tag{14}$$

$$A = \int \phi \cdot u dxdydz , \tag{15}$$

where

$$(u_t)^2 = \left(\frac{\partial u_1}{\partial t}\right)^2 + \left(\frac{\partial u_2}{\partial t}\right)^2 + \left(\frac{\partial u_3}{\partial t}\right)^2 ,$$

$$\phi \cdot u = Xu_1 + Yu_2 + Zu_3 .$$

By virtue of Hamilton's principle, the shift vector u of points of the elastic body D being in stress state gives a stationary value for the integral

$$E(u) = \int_{t_1}^{t_2} (E_d - E_k - A)dt . \tag{16}$$

Based on (13), (14) and (15) it is easy to see that when ρ is independent of t, the Euler-Lagrange equations of the functional (16) can take the form

$$\rho u_{tt} - G\left(\Delta u + \frac{m}{m-2} grad\ divu\right) - \phi = 0 . \tag{17}$$

The vector equality (17) is a system of *partial differential equations for the linear elasticity theory*, describing shift components.

By the notations $\lambda = \frac{2G}{m-2}$ and $\mu = G$ the system (17) is usually written in the form

$$\rho u_{tt} - \mu\Delta u - (\lambda + \mu) grad\ divu - \phi = 0 . \tag{18}$$

When an elastic body is situated in a static (equilibrium) stress state, i.e. when the shift vector μ is a function of spatial variables only, instead of (18),

$$\mu\Delta u + (\lambda + \mu) grad\ div\ u = 0 \tag{19}$$

is valid due to absence of volume forces.

The equalities expressed by formula (19) are called the *system of equations for static linear theory of elasticity* with respect to shift components u_1, u_2, u_3.

In general, we say that an elastic body is subjected to the planar deformation parallel to the plane of the variables x, y, if the third, u_3, among three shift components u_1, u_2 and u_3 is equal to zero and the first two components, $u_1 = u$ and $u_2 = v$, are functions of x, y only.

In this case, for determining u and v we have from (19) the system of equations

$$\begin{aligned} \mu\Delta u + (\lambda + \mu)(u_{xx} + v_{xy}) = 0 , \\ \mu\Delta v + (\lambda + \mu)(u_{xy} + v_{yy}) = 0 . \end{aligned} \tag{20}$$

For given shift components u_1, u_2 and u_3, the formulae (12) make it possible to determine the deformation components.

Static theory of elasticity has as its subject the investigation of static stress state for elastic bodies if the law of dependency among the stress components and deformation components is known. In the linear theory of elasticity this law is the so-called *generalized Hooke's law*, according to which the stress components at every point of an elastic body are homogeneous linear functions (linear forms) of the deformation components, and vice versa. If the coefficients of this linear relation are constants, thc clastic body is called *homogeneous*.

1.5.4. For homogeneous elastic bodies in the case of planar deformation, the generalized Hooke's law claims that

$$X_x = \lambda(e_{xx} + e_{yy}) + 2\mu e_{xx}\ , \quad Y_y = \lambda(e_{xx} + e_{yy}) + 2\mu e_{yy}\ ,$$
$$X_y = Y_x = \mu e_{xy}\ , \quad Z_z = \lambda(e_{xx} + e_{yy})\ , \quad X_z = Z_x = Y_z = Z_y = 0\ .$$

By virtue of (12) we can add to these equalities the following,

$$\begin{gathered} X_x = \lambda(u_x + v_y) + 2\mu u_x\ , \quad Y_y = \lambda(u_x + v_y) + 2\mu v_y\ , \\ X_y = Y_x = \mu(v_x + u_y)\ , \quad Z_z = \lambda(u_x + v_y)\ , \\ X_z = Z_x = Y_z = Z_y = 0\ . \end{gathered} \tag{21}$$

Excluding u and v from (21), we find, based on (20), that

$$\frac{\partial X_x}{\partial x} + \frac{\partial X_y}{\partial y} = 0\ , \quad \frac{\partial Y_x}{\partial x} + \frac{\partial Y_y}{\partial y} = 0\ . \tag{22}$$

When a continous medium is in a static stress state, the principal vector and principal moment of all forces acting on it should be equal to zero. The fulfilment of the first of these requirements gives three equations concerning equilibrium:

$$\begin{gathered} \frac{\partial X_x}{\partial x} + \frac{\partial X_y}{\partial y} + \frac{\partial X_z}{\partial z} + X = 0\ , \quad \frac{\partial Y_x}{\partial x} + \frac{\partial Y_y}{\partial y} + \frac{\partial Y_z}{\partial z} + Y = 0 \\ \frac{\partial Z_x}{\partial x} + \frac{\partial Z_y}{\partial y} + \frac{\partial Z_z}{\partial z} + Z = 0\ , \end{gathered} \tag{23}$$

and the fulfilment of the second requirement shows $X_y = Y_x$, $X_z = Z_x, Y_z = Z_y$. The latter equalities mean that stress tensor is symmetric.

In case of planar deformation, the system (22) is obtained from the system (23) under the condition of absence of volume forces. Thus, in order to determine the five variables $X_x, X_y = Y_x, Y_y$, u and v which characterize the planar stress state of an elastic body, we have a system of five equations, namely the first three equations of (21) and the two equations of (22). We obtain the system (20) from this system as a result of eliminating X_x, X_y, Y_y.

§1.6 Basic Equations of Fluid Mechanics

1.6.1. On every conceptually chosen surface σ in a volume of liquid, pressure acts. When pressure directedly acts along the normal onto an area element of σ, in general the law of isotropy for pressure is valid. This law is always valid if the liquid is in an equilibrium state. If this stated law is also valid for some moving liquid, then that liquid is called *ideal.*

In a liquid we extract an arbitrary small volume τ—"particle of liquid". Some specific mean velocity, the limit of which as τ contracts and tends to a point $P \in \tau$ is called the velocity q of the liquid at the point $P \in \tau$ is ascribed to the liquid filling the volume τ. It is clearly a vector. We denote by x_1, x_2, x_3 the orthogonal Cartesian coordinates of point P, and by q_1, q_2, q_3 the components of the vector q at point P. The motion or equilibrium state of the liquid is determined if the pressure p, the distribution density of mass ρ, and the vector of velocity q at every point $P(x, y, z)$ are known.

A curve in the volume occupied by a liquid is called a *streamline* if its tangent at every point is parallel to the velocity vector at that point. If the velocity vector is independent of time at every point of a streamline, then this curve coincides with the trajectory of a moving point which has as its equation of motion

$$\frac{dx}{dt} = q ,$$

where x is the radius vector of point P.

If the density ρ is constant for all the volume occupied by the liquid, the liquid is called *incompressible.*

In the presence of friction (interior, between particles of liquids or, exterior, between particles of liquid and rigid bodies which are in contact with the particles), the isotropic law of pressure for moving fluid, may fail to hold, in which case the fluid is called *real* or *actual.*

Interior friction is the reason for the appearance of the resistance and the buoyant force in back of the curl for a rigid body in a real fluid. They vanish at once in the equilibrium state. This fact is found to be closely connected with the fact that interior friction can appear in the case the change of fluid form is resisted. When we study motion of fluid, if we cannot ignore the interior friction then the fluid is called *viscous*.

If the velocity q of a moving fluid at every point is independent of time, then the motion is called *stationary*.

1.6.2. In Hydromechanics and in very general cases when we suppose that volume forces and thermal phenomena are absent, then in order to determine variables q, p and ρ it is very important that the system of equations which consist of the *Navier-Stokes equation*

$$\rho \frac{Dq}{Dt} + grad\ p = \mu \Delta q \tag{24}$$

and the scalar *continuity equation*

$$\frac{\partial \rho}{\partial t} + div\ \rho q = 0 \tag{25}$$

where differential operators *grad*, *div* and $\Delta = div\ grad$ are taken with respect to spatial variables $x_1, \dots, x_n$, and $\frac{D}{Dt}$ is the so called *substantial derivative* with respect to time

$$\frac{D}{Dt} = \frac{\partial}{\partial t} + q\ grad\ , \quad q\ grad = q\frac{\partial}{\partial x_1} + q_2 \frac{\partial}{\partial x_2} + q_3 \frac{\partial}{\partial x_3}\ , \tag{26}$$

and μ is the coefficient of viscosity.

Finally, the fifth equation of state

$$p = p(\rho) \tag{27}$$

is also added to the system (24), (25) of four equations.

In the case of the so-called barotropic fluid the relation (27) is monoton.

If we denote by $q^2 = q \cdot q$ the scalar product of the vector q with itself, and by $\Omega \times q$ the vector product of the vector $\Omega = rot\ q$ and q, then by virtue of (26) the equation (24) can be put the form

$$\rho \frac{\partial q}{\partial t} + p \left[\frac{1}{2} grad\ q^2 + \Omega \times q \right] + grad\ p = \mu \Delta q\ . \tag{28}$$

The vector Ω is called a *curl* or rotation of the velocity q. The motion of fluid is called *rotational* or *irrotational* depending on $\Omega \neq 0$ or $\Omega = 0$ everywhere in the volume occupied by the fluid.

In the case of stationary motion of a fluid Eqs. (28), (25) reduce to

$$\rho\left[\frac{1}{2} grad\, q^2 + \Omega \times q\right] + grad\, \rho = \mu \Delta\, q\,, \tag{29}$$

and

$$div\, \rho\, q = 0\,, \tag{30}$$

respectively.

For irrotational motions the absence of the term $\rho\Omega \times q$ in Eq. (28) is characteristic. In the absence of rotation and viscosity,

$$\frac{1}{2} grad\, q^2 + \frac{1}{\rho} grad\, \rho = 0$$

holds instead of (29).

1.6.3. The system of equations (29) and (30) is nonlinear. If the value of $grad\, q^2 + 2\Omega \times q$ is small enough so that it can be neglected, then in the study of stationary motions we can confine ourselves to considering the linear system

$$\Delta\, q - \frac{1}{\mu}\, grad\, p = 0\,, \quad div\, q = 0\,.$$

Notice that the equation of continuity (25) is a differential representation of the assumption that in a moving fluid there are neither source nor sink, i.e., in the volume occupied by the fluid neither a loss nor a rise of its mass takes place. As to Eq. (28), we can derive it easily in a similar way from *d'Alembert's principle*, in accordance with which at every instant of the motions of an arbitrary continuous medium all forces applied to it—volume force are in balance with each other.

If there exists a family of parallel planes on which the hydromechanical situation is the same where each intersects the volume occupied by the fluid, then that motion is called *plane-parallel*. It can be considered as a two-component motion by selecting a coordinate system, because in this case the vector velocity q is parallel to the place of motion.

§1.7 Equations in General Theory of Relativity

1.7.1. The study of the universal property of interaction of bodies which is called *gravitation*. It is one of the central problems in modern physics, and has general scientific and world view significance.

In Newton's theory of gravitation the coordinates of three-dimensional Euclidean space are taken as parameters for determining the position of a mass point, and the geometric space and the time are present independently of each other. In accordance with this theory, a mass distributed in some volume with a density ρ acts on the point mass m with a force f which is a three-component vector determined by the formula

$$f = m \; grad \; \varphi \; ,$$

where the scalar function φ, the Newtonian potential, is the solution of Poisson's equation

$$\Delta\varphi = 4\pi\gamma\rho \; , \tag{31}$$

where Δ is the Laplacian in spatial variables and γ is the gravitation constant.

Newton's theory of gravitation plays an essential role in physics, natural sciences and technology, but it cannot explain a number of actually observed phenomena like, for example the shift of the perihelion of Mercury, the distortion of light paths near the sun, red shift, etc. This has become the cause of the advent of a new theory of gravitation—the general theory of relativity.

1.7.2. According to general theory of relativity the parameters x_1, x_2, x_3, x_4 which characterize the motion of bodies are the coordinates of a point in 4-dimensional Reimann space R_4, the so-called world space, with the metric quadratic form

$$ds^2 = \sum_{\alpha,\beta=1}^{4} g_{\alpha\beta} \; dx_\alpha dx_\beta \; ,$$

where the coefficients $g_{\alpha\beta}$ satisfy the system of *Einstein's equations*

$$R_{\mu\nu} - \frac{1}{2} g_{\mu\nu} R = -\kappa T_{\mu\nu} \; , \quad \mu, \nu = 1, 2, 3, 4 \; , \tag{32}$$

where $R_{\mu\mu}$ is an abridged curvature tensor:

$$R_{\mu\nu} = \frac{\partial}{\partial x_\nu}\Gamma^\alpha_{\alpha\mu} - \frac{\partial}{\partial x_\alpha}\Gamma^\alpha_{\mu\nu} + \Gamma^\alpha_{\beta\nu}\Gamma^\beta_{\alpha\mu} - \Gamma^\alpha_{\alpha\beta}\Gamma^\beta_{\mu\nu} \tag{33}$$

with

$$\Gamma^\alpha_{\mu\nu} = -\frac{1}{2}g_{\mu\beta}\frac{\partial g^{\alpha\beta}}{\partial x_\nu} - \frac{1}{2}g_{\nu\beta}\frac{\partial g^{\alpha\beta}}{\partial x_\mu} - \frac{1}{2}g^{\alpha\beta}\frac{\partial g_{\mu\nu}}{\partial x_\beta}\,, \tag{34}$$

R is the invariant of tensor $R_{\mu\nu}$:

$$R = g^{\alpha\beta}\frac{\partial^2 \log\sqrt{g}}{\partial x_\alpha \partial x_\beta} + \frac{1}{\sqrt{-g}}\frac{\partial^2 g^{\alpha\beta}}{\partial x_\alpha \partial x_\beta} + \frac{1}{2}\Gamma^\nu_{\alpha\beta}\frac{\partial g^{\alpha\beta}}{\partial x_\nu}\,,$$

$T_{\mu\nu}$ is the material tensor, and κ is a proportional coefficient. The repetition of an index in the formulae (33) and (34), as well as in the expression of R, denotes that the summation in this index is taken from one to four. Same as in §2, every element of the matrix $\|g^{\alpha\beta}\|$ is equal to the ratio of the cofactor of the corresponding element in the matrix $\|g_{ij}\|$ to the determinant $g = \det\|g_{ij}\|$. The variables $g_{\alpha\beta}$ and $g^{\alpha\beta}$ are usually called, respectively, ***covariant*** and ***contravariant components*** of the fundamental tensor.

In general theory of relativity a significant difficulty appears when we search for an expression of the material tensor $T_{\mu\nu}$. Naturally, all components of $T_{\mu\nu}$ are equal to zero in the part of the world space where material does not exist. Because matter is distributed non-uniformly in the world space, and, moreover, it concentrates in the form of moving bodies so that their dimensions are small compared with the distances among them, for a number of cases in theory of gravitation it is justified to use an alternative version of Eq. (32) in which all components of the material tensor are equal to zero, i.e.,

$$R_{\mu\nu} - \frac{1}{2}g_{\mu\nu}R = 0\,. \tag{35}$$

From the notations of (33) and (34) it follows that (35) is a system of nonlinear partial differential equations with respect to $g_{\alpha\beta}$. A knowledge of the functions $g_{\alpha\beta}$ makes it possible to determine the material properties of world space, and, moreover, it is especially important that the motion of a mass point be defined as a distribution of singularity of these functions.

§1.8 Model Partial Differential Equations

1.8.1. Partial differential equations obtained from investigating some specific phenomena can prove useful also for some other phenomena. This can be easily illustrated using the simplest of the so-called model equations like, for example, the *wave equation*

$$\frac{\partial^2 u}{\partial t^2} - \Delta u = 0 \ , \tag{36}$$

the *heat-conduction equation*

$$\frac{\partial u}{\partial t} - \Delta u = 0 \ , \tag{37}$$

and the *Laplace equation*

$$\Delta u = 0 \ , \tag{38}$$

where $\Delta = \sum_{i=1}^{n} \frac{\partial^2}{\partial x_i^2}$ is the Laplacian.

As shown, in §3, the Eq. (36) for $n = 2, x_1 = x$ and $x_2 = y$ describes the *oscillation of a membrane*. Equation (36) for $n = 1$ and $x_1 = x$ coincides with the *oscillating string equation*

$$\frac{\partial^2 u}{\partial t^2} - \frac{\partial^2 u}{\partial x^2} = 0 \ . \tag{39}$$

As it is well-known Eq. (39) can also be applied to the investigation of electric fluctuation in wires.

We now show that Eq. (36) with $n = 3$ models the process of sound propagation in a gas. In fact, we can make use of the equations of hydromechanics Eq. (25), (27) and (28), because for this process the gas performs oscillating motion as its characteristics. We suppose that (1) the state equation (27) has the form

$$\frac{p}{p_0} = \left(\frac{\rho}{\rho_0}\right)^{\gamma} \ , \quad \gamma = \frac{c_p}{c_v} \ , \tag{40}$$

where p_0 and ρ_0 are the values of the pressure p and mass density ρ of the gas at the starting point of time, and c_p, c_v are the heat capacities under the conditions of constant pressure and constant volume respectively; (2) *viscosity* is absent; (3) the velocity q, its first derivatives, the variation $1 - \frac{\rho}{\rho_0}$ of density ρ and its derivatives are small enough so that we can neglect the following quantitives:

$$grad\, q^2 \ , \quad \Omega \times q \ , \quad (\rho - \rho_0)\, div\, q \ , \quad q\, grad\, \rho \ ,$$

$$\left(1 - \frac{\rho}{\rho_0}\right)^k grad\, p \ , \quad k \geq 1, \quad I\left(1 - \frac{\rho}{\rho_0}\right)^k, \quad k > 1 \ .$$

By virtue of

$$div\ pq = q\ grad\ \rho + \rho\ div\ q\ ,$$
$$\frac{1}{\rho} = \frac{1}{\rho_0}\sum_{k=0}^{\infty}\left(1 - \frac{\rho}{\rho_0}\right)^k\ ,$$
$$\left(\frac{\rho}{\rho_0}\right)^{\gamma} = \left(1 + \frac{\rho - \rho_0}{\rho_0}\right)^{\gamma}$$

applying the above assumptions, (1), (2) and (3); Eqs. (25), (28) and (40) can be replaced respectively by the equations

$$\frac{\partial\rho}{\partial t} = -\rho_0\ div\ q\ , \tag{41}$$
$$\frac{\partial q}{\partial t} = -\frac{1}{\rho_0} grad\ p \tag{42}$$

and

$$p = p_0(1-\gamma) + \frac{p_0}{\rho_0}\gamma\rho\ . \tag{43}$$

On account of (43), Eq. (42) has the form

$$\frac{\partial q}{\partial t} = \frac{\gamma p_0}{\rho_0^2} grad\ \rho\ . \tag{44}$$

Differentiating Eq. (41) with respect to t and subjecting Eq. (44) to the operator div, we have

$$\frac{\partial^2\rho}{\partial t^2} = -\rho_0\ div\frac{\partial q}{\partial t}\ , \quad div\frac{\partial q}{\partial t} = -\frac{\gamma p_0}{\rho_0^2}\Delta\rho\ .$$

From here we arrive at the conclusion that ρ is a solution of equation (7) with $a^2 = \frac{\gamma p_0}{\rho_0}$, and thus the stated assertion has been proved.

In the various physical phenomena mentioned above, the continuous medium does oscillating motion. Therefore these phenomena are naturally modelled mathematically in terms of the wave equation (36) with certain acceptable accuracy.

1.8.2. The equation (37) can also be used to describe the so-called *transport phenomena*, in addition to modelling heat propagation (cf. §1.4). For instance, the concentration of matter in an irregular distribution, called a *diffusion*, under certain assumptions is described by equation

$$c\frac{\partial u}{\partial t} - D\Delta u = 0 , \tag{45}$$

where c is the porosity coefficient, D is the diffusion coefficient, and $u(x,t)$ is the concentration of matter at point x in the medium at the instant t. When c and D are constants (they are both positive), Eq. (45) reduces to Eq. (37).

1.8.3. Laplace's equation (38) for $n = 2, x_1 = x, x_2 = y$, as we had already noticed at the end of §3, describes the equilibrium state of a membrane with deflection $u = u(x,y)$. From the formula (31) it follows that Newtonian mass potential is a solution of Laplace's equation in those domains of gravity field where mass is absent.

Equation (38) for $n = 2$ and $n = 3$ plays an important role in studying *electric field.* We shall confine ourselves to considering a planar electrostatic field, i.e., a planar medium in which a two-component vector $\mathbf{E} = (E_x, E_y)$—*field force*—is defined for every point $P(x,y)$. Let D be an arbitrary domain of the field with smooth boundary $S = \partial D$. The expressions

$$N = \int_s E \cdot \nu \, ds \ , \quad A = \int_s E \cdot s \, ds \ ,$$

where ν and s are respectively the unit outward normal vector and unit tangent vector of s with positive direction at the integrated point, are called the flow across the contour s and the circulation along S and the vector E, respectively.

By the Gauss-Ostrogradskii formula we have

$$N = \int_D \left(\frac{\partial E_x}{\partial x} + \frac{\partial E_y}{\partial y} \right) dxdy \ , \quad A = \int_D \left(\frac{\partial E_y}{\partial x} - \frac{\partial E_x}{\partial y} \right) dxdy \ .$$

Contracting, the domain D to a point P, we obtain as the limit

$$\lim_{D \to P} \frac{N}{|D|} = \frac{\partial E_x}{\partial x} + \frac{\partial E_y}{\partial y} = div \ E \ ,$$

$$\lim_{D \to P} \frac{A}{|D|} = \frac{\partial E_y}{\partial x} - \frac{\partial E_x}{\partial y} = rot \ E \ .$$

As $N = 4\pi e$ and $\lim_{D \to P} \frac{N}{|D|} = 4\pi\rho$, where e is the total change situated in D and ρ the surface density of charge at point P, we can write

$$\frac{\partial E_x}{\partial x} + \frac{\partial E_y}{\partial y} = 4\pi\rho \ . \tag{46}$$

Becasue A is the work of the force $\mathbf{E}$ on path S and the field is static, $A = 0$ by virtue of the conservation law of energy, thus

$$\frac{\partial E_y}{\partial x} - \frac{\partial E_x}{\partial y} = 0 . \tag{47}$$

The fulfilment of equality (47) means that the expression $-E_x dx - E_y dy = dv$ is a total differential.

Hence $\frac{\partial v}{\partial x} = -E_x$, $\frac{\partial v}{\partial y} = -E_y$ for the function $v(x, y)$ which is called *potential of the field*, and from (46) we obtain the equation

$$\Delta v = -4\pi\rho . \tag{48}$$

The presence of the function v makes it possible to write an electric field in terms of density. From (48) it follows that the field potential satisfies Laplace's equation everywhere where there are no changes.

1.8.4. In the cases the phenomena under consideration are extremely nonlinear the partial differential equations to model them are also nonlinear. It is usually useless, and should not be considered as correct, to ignore these nonlinearity without an exposition of their characters. Therefore, we have to replace complex nonlinear models with simpler, but also nonlinear, models. In the last ten years specialists have conducted interesting investigations into these models of nonlinear partial differential equations which are important for applications such as, for example, *sine-Gordon's equation*

$$u_{tt} - u_{xx} = -m \sin u , \quad m = \text{const} ,$$

Schrödinger's equation

$$iu_t + u_{xx} + 2|u|^2 u = 0 ,$$

Hopf-Cale's equation

$$u_{xx} - \frac{1}{2\nu} u_x^2 - u_t = 0 , \quad \nu = \text{const} ,$$

Kortjeveg-de Friz's equation

$$u_t + 6uu_x + u_{xxx} = 0 ,$$

Monge-Ampere's equation

$$u_{xx} u_{yy} - (u_{xy})^2 = a , \quad a = \text{const}$$

and so on.

Chapter II. Basic Types of Partial Differential Equations and Wellposedness of Statement of the Problems for Them

§2.1 Concept of Characteristic Form and Type Division of Partial Differential Equations

2.1.1. As already noted at the end of §1.1, chapter I, we can, without loss of generality, suppose that all the variables entering the equation

$$F\left(x,\cdots,\frac{\partial^k u}{\partial x_1^{i_1}\cdots\partial x_n^{i_n}}\cdots\right)=0\ ,\quad k=\sum_{l=1}^{n} i_l \tag{1}$$

are real.

Suppose that the number N of the equations in the vector equality (1) is equal to the number N_1 of independent functions, $N=N_1$, and that the order of every one of these equations is equal to m with, moreover, $m>1$ if $N=1$.

When partial-derivatives of the functions $F_1,\cdots,F_N$ are continuous with respect to all

$$p^j_{i_1\cdots i_n}=\frac{\partial^m u_j}{\partial x_1^{i_1}\cdots\partial x_n^{i_n}}\ ,\quad \sum_{l=1}^{n} i_l=m\ ,$$

the linear parts of the increments of these functions

$$\sum_{i_1\cdots i_n}\sum_{j=1}^{N}\frac{\partial F_i}{\partial p^j_{i_1\cdots i_n}}dp^j_{i_1\dots i_n}\ ,\quad i=1,\cdots,N$$

are principal. Therefore, they naturally play an important role in the theory of partial differential equations of the form (1).

With the help of the square matrix

$$\left\|\frac{\partial F_i}{\partial p^j_{j_1\cdots i_n}}\right\|\ ,\quad i,j=1,\cdots,N\ ,\quad \sum_{l=1}^{n} i_1=m\ ,$$

we constitute an Nm-order form in real parameters $\lambda_1,\cdots,\lambda_n$,

$$K(\lambda_1,\cdots,\lambda_n)=\det\sum_{i_1,\cdots,i_n}\left\|\frac{\partial F_i}{\partial p^j_{i_1\cdots i_n}}\right\|\lambda_1^{i_1}\cdots\lambda_n^{i_n}\ , \tag{2}$$

where the summation is taken on an index set of nonnegative integers $i_1, \cdots, i_n$ that satisfy the condition $\sum_{s=1}^{n} i_s = m$.

The expression (2), in which

$$p^j_{i_1 \cdots i_n} = \frac{\partial^k u_j}{\partial x_1^{i_1} \cdots \partial x_n^{i_n}}, \quad j = 1, \cdots, N, k = 0, \cdots, m,$$

is called a *characteristic form (characteristic determinant)* of the equation (1). When Eq. (1) is linear, the coefficients of the form (12) depend only on point x, whereas in the general case they are also functions of the unknown solution $U(x)$ and its derivatives.

2.1.2. In accordance with characteristic form (2), partial differential equations of the form (1) are divided into various types.

We first assume that equation (1) is linear. If, for a fixed point x in domain D, we can find a non-singular affine transformation of arguments

$$\lambda_i = \lambda_i(\mu_1, \cdots, \mu_n), \quad i = 1, 2, \cdots, n,$$

such that the form resulting from (2) contains only l arguments $\mu_i, 0 < l < n$, then we say that Eq. (1) degenerates *parabolically*.

If in the space of the arguments $\lambda_1, \cdots, \lambda_n$ the conic manifold

$$K(\lambda_1, \cdots, \lambda_n) = 0 \tag{3}$$

has no real point, except for $\lambda_1 = \lambda_2 = \cdots = \lambda_n = 0$, then eq. (1) is called *elliptic* at point x. When Eq. (1) does not have parabolic degeneration, we say that equation (1) is *hyperbolic* at point x if in the space of the arguments $\lambda_1, \cdots, \lambda_n$ there exists such a straight line that, as long as we take it as a coordinate axis of the new arguments $\mu_1, \cdots \mu_n$ obtained by a non-singular affine transformation of $\lambda_1, \cdots, \lambda_n$, then with respect to the coordinate changing along this line, the transformed equation (3) has exactly N real (simple or multiple) roots for any selection of the values for the other μ coordinates.

2.1.3. Similarly in the nonlinear case Eq. (1) is classified into types according to the character of the form (2) in which $p^j_{i_1 \cdots i_n}$ are replaced by $\frac{\partial^k u_j}{\partial x_1^{i_1} \cdots \partial x_n^{i_n}}, \sum_{s=1}^{n} i_s = k, k = 0, \cdots, m$. As this time the coefficients of the form (2) depend on the unknown solution and its derivatives in addition to point

x, the classification into types in the case considered makes sense only for this solution.

§2.2 Classification into Types of Linear Second Order Partial Differential Equations

2.2.1. Without loss of generality, suppose that the linear partial differential equation of second order is written in the form (cf. Chapter 1, §1.1.)

$$Lu \equiv \sum_{i,j=1}^{n} A_{ij}(x)\frac{\partial^2 u}{\partial x_i \partial x_j} + \sum_{i=1}^{n} B_i(x)\frac{\partial u}{\partial x_i} + C(x)u = f(x) \; . \tag{4}$$

When (4) is a system of equations, it is assumed that A_{ij}, B_i and C are $N \times N$ square matrices, and f, u are N-dimensional vectors. Moreover, if $G = \|G_{ik}\|$ is a square $N \times N$ matrix and $v = (v_1, \cdots, v_n)$ being a vector with N components, then by the product Gv we mean the N-dimensional vector with components

$$(Gv)_l = \sum_{k=1}^{n} G_{lk} v_k \; , \quad l = 1, \cdots, N \; .$$

Whenever $N = 1$ the characteristic form (2) that corresponds to equation (4) is the *quadratic form*

$$K(\lambda_1, \cdots, \lambda_n) \equiv Q(\lambda_1, \cdots, \lambda_n) = \sum_{i,j=1}^{n} A_{ij}(x)\lambda_i \lambda_j \; . \tag{5}$$

2.2.2. In this and the next section, by (4) we mean a scalar equation.

In accordance with the definition given in the previous section Eq. (4) at a point x (assuming that the degeneration of order for the equation is excluded) will be elliptic, hyperbolic or parabolic depending on whether the form (5) is definite (positive or negative), with alternating signs or degenerated.

At every point x the quadratic form Q can be reduced to a canonical form $Q = \sum\limits_{i=1}^{n} \alpha_i \mu_i^2$ with the help of some nonsingular affine transformation $\lambda_i = \lambda_i(\mu_1, \cdots, \mu_n), i = 1, \cdots, n$, where $\alpha_i, i = 1, 2, \cdots, n$, take values 1, -1 and 0, and the number of negative coefficients (the inertia index) and the number of zero coefficients (deficiency index) are affinely invariant.

If all $\alpha_i = 1$ or all $\alpha_i = -1, i = 1, \cdots, n$, i.e., if the form Q is correspondingly positive definite or negative definite, Eq. (4) is elliptic at point x. If one of the coefficients α_i is negative while all the remaining ones are positive (or vice versa), then Eq. (4) at point x is hyperbolic. In the case that with respect to some l, $1 < l < n-1$, the coefficients $\alpha_j, i \leq l$, are all positive and the $n-l$ of the remaining ones are negative, then the equation is called *ultra-hyperbolic.* If at least one of these coefficients is equal to zero (α_i cannot all be zeros since the degeneration of order for the equation is excluded), then Eq. (4) is parabolic at point x.

We say that the equation (4) in its domain of specification is of the elliptic, hyperbolic or parabolic type if it is correspondingly elliptic, hyperbolic or parabolic at every point of this domain.

Equation (4) elliptic in domain D is called *uniformly elliptic*, if there exist real numbers k_0 and K_1 which are different from zero and are of the same sign, such that for all $x \in D$

$$K_0 \sum_{i=1}^{n} \lambda_i^2 \leq Q(\lambda_1, \cdots, \lambda_n) \leq k_1 \sum_{i=1}^{n} \lambda_i^2 .$$

As an example, the equation

$$x_n \frac{\partial^2 u}{\partial x_1^2} + \sum_{i=2}^{n} \frac{\partial^2 u}{\partial x_i^2} = 0 \tag{6}$$

shows that an equation which is elliptic in its domain of specification may not be uniformly elliptic. Equation (6) is elliptic in every point in the half-space $x_n > 0$, but is not uniformly elliptic in it.

If Eq. (4) belongs to different types in different parts of domain D, we say that it is *a mixed type equation* in this domain. Equation (6) gives us an example of a mixed type equation in any domain D of the space E_n where its intersection with the hyperplane $x_n = 0$ is not empty.

In what follows, by definiteness of a quadratic form we mean positive definiteness, since a negative definite quadratic form can be multiplied by (-1) to become positive definite.

Assuming, without loss of generality, that the form Q is symmetric, i.e. $A_{ij} = A_{ji}, i, j = 1, \cdots, n$, and using Sylvester's criteria for positive definiteness of a quadratic form, we can, without reducing the quadratic form Q to canonical form at every point $x \in D$, claim that the necessary and sufficient condition for ellipticity of Eq. (4) in domain D is that all principal diagonal minors of the matrix

$$A = \|A_{ij}\|$$

are positive in all points of this domain.

At every fixed point $x \in D$ we can always find a nonsingular transformation (exchange) $y = y(x)$ of independent variables $y_i = y_i(x_1, \cdots, x_n), i = 1, \cdots, n$, by which Eq. (4) is carried into a canonical form

$$\sum_{i=1}^{n} \left(\alpha_i \frac{\partial^2 v}{\partial y_i^2} + \beta_i \frac{\partial v}{\partial y_i} \right) + \gamma v = \sigma \ , \tag{7}$$

where the constants $\alpha_i, i = 1, \cdots, n$, take values 1, -1, 0,

$$v(y) = u[x(y)] \ , \quad \sigma(y) = f[x(y)] \ ,$$

and the functions β_i, γ are expressed in terms of the coefficients of Eq. (4).

Indeed, we take as that transformation the affine transformation

$$y_i = \sum_{k=1}^{n} g_{ki} x_k \ , \quad i = 1, \cdots, n \ ,$$

where $\|g_{ik}\|$ is the matrix of nonsingular affine transformation $\lambda_i = \sum\limits_{k=1}^{n} g_{ik} \mu_k$, $i = 1, \cdots, n$, which reduces the form (5) into its canonical form $Q = \sum\limits_{s=1}^{n} \alpha_s \mu_s^2$, that is,

$$Q = \sum_{k,l=1}^{n} a_{kl} \mu_k \mu_l \ , \quad a_{kl} = \sum_{s=1}^{n} A_{ij} g_{ik} g_{jl} \ ,$$

$$a_{kl} = \begin{cases} \alpha_s \ , & k = l = s \ , \\ 0 \ , & k \neq l \ . \end{cases}$$

From here and taking into account

$$\frac{\partial}{\partial x_i} = \sum_{k=1}^{n} g_{ik} \frac{\partial}{\partial y_k} \ , \quad \frac{\partial^2}{\partial x_i \partial x_j} = \sum_{k,l=1}^{n} g_{ik} g_{jl} \frac{\partial^2}{\partial y_k \partial y_l}$$

and

$$\sum_{i,j=1}^{n} A_{ij} \frac{\partial^2 u}{\partial x_i \partial x_j} = \sum_{i,j=1}^{n} A_{ij} \sum_{k,l=1}^{n} g_{ik} g_{jl} \frac{\partial^2 v}{\partial y_k \partial y_l}$$

$$= \sum_{k,l=1}^{n} a_{kl} \frac{\partial^2 v}{\partial y_k \partial y_l} = \sum_{i=1}^{n} \alpha_i \frac{\partial^2 v}{\partial y_i^2} ,$$

$$\sum_{i=1}^{n} B_i \frac{\partial u}{\partial x_i} = \sum_{i=1}^{n} B_i \sum_{k=1}^{n} g_{ik} \frac{\partial v}{\partial y_k} = \sum_{k=1}^{n} \beta_k \frac{\partial v}{\partial y_k} ,$$

$$\beta_k = \sum_{i=1}^{n} B_i g_{ik} , \quad k = 1, \cdots, n ,$$

we arrive at the equality (7) at once.

2.2.3. It should be noted that it is not always possible to find a transformation of independent variables which allows us to reduce Eq. (4) to the canonical form (7) not only in all of the domain of specification of this equation but even near the given point of the domain. In this respect the case of two independent variables is an exception.

We write Eq. (4) for $n = 2$ with the notations

$$x_1 = x , \quad x_2 = y ,$$
$$A_{11} = a(x,y) , \quad A_{12} = A_{21} = b(x,y) , \quad A_{22} = c(x,y) ,$$
$$B_1 = d(x,y) , \quad B_2 = e(x,y) , C = g(x,y) , \quad f = f(x,y)$$

in the form

$$au_{xx} + 2bu_{xy} + cu_{yy} + du_x + eu_y + gu = f . \tag{8}$$

The curve $\varphi(x,y) = $ constant, where $\varphi(x,y)$ is the solution of the equation

$$a\varphi_x^2 + 2b\varphi_x\varphi_y + c\varphi_y^2 = 0 , \tag{9}$$

is called a *characteristic curve* of Eq. (8), and the direction (dx, dy) defined by

$$ady^2 - 2bdxdy + cdx^2 = 0 \tag{10}$$

is called a *characteristic direction*.

The Equation (10) is the ordinary differential equation of the characteristic curve.

According to the definition introduced above, Eq. (8) is elliptic, hyperbolic or parabolic depending on whether the quadratic form $ady^2 + 2bdxdy + cdx^2$ is definite (positive or negative), with alternating signs or semi-definite (degenerate). Corresponding to these Eq. (8) is elliptic, hyperbolic or parabolic depending on whether the discriminant $ac - b^2$ of the quadratic form $ady^2 + 2bdxdy + cdx^2$ is more than zero, less than zero or equal to zero, respectively. Therefore, rewriting the differentail equation of characteristic curve (10) in the form

$$ady = \left(b \pm \sqrt{b^2 - ac}\right) dx ,$$

we arrive at the conclusion that Eq. (8) has no real characteristic direction in domain of ellipticity, has two different characteristic directions at each point of hyperbolicity, and has one real characteristic direction at each point of parabolicity. Besides, it is clear that in the domain of ellipticity Eq. (8) has no real characteristic curves; but, under the condition of sufficiently smooth coefficients a, b and c, the domain of hyperbolicity of Eq. (8) is covered by a net of two families of characteristic curves, and the domain of parabolicity is covered by one family of characteristic curves.

In the case of

$$y^m u_{xx} + u_{yy} = 0 \tag{11}$$

where m is an odd natural number, equation (10) has the form

$$y^m dy^2 + dx^2 = 0 . \tag{12}$$

From this it follows that this equation has no real characteristic directions in the half-plane $y > 0$, while at each point of the straight line $y = 0$ and the half-plane $y < 0$ it has respectively one characteristic direction and two characteristic directions. Rewriting the differential equation (12) of characteristic curves in the form $dx \pm (-y)^{\frac{m}{2}} dy = 0$ and integrating it, we conclude that the half-plane $y < 0$ is covered by the two families of characteristic curves of equation (11),

$$x - \frac{2}{m+2}(-y)^{\frac{m+2}{2}} = \text{const.} ,$$

and

$$x + \frac{2}{m+2}(-y)^{\frac{m+2}{2}} = \text{const.} .$$

In an arbitrary domain D in the plane of variables x, y its intersection with the straight line $y = 0$ is not empty, (11) is a mixed type equation: it is elliptic whenever $y > 0$, hyperbolic whenever $y < 0$ and degenerate parabolicaly whenever $y = 0$.

§2.3 Canonical Form of Linear Second Order Partial Differential Equations with Two Independent Variables

2.3.1. Suppose that a linear second order partial differential equation with two independent variables has been written in the form (8). Under the condition that the coefficients a, b, c are sufficiently smooth, a nonsingular transformation of independent variables $x, y, \xi = \xi(x, y)$ and $\eta = \eta(x, y)$, can always be found such that Eq. (8) in its domain of specification D is reduced to one of the following canonical forms:

$$v_{\xi\xi} + v_{\eta\eta} + Av_{\xi} + Bv_{\eta} + Cv = f_0 \tag{13_1}$$

in elliptic case,

$$v_{\xi\eta} + Av_{\xi} + Bv_{\eta} + Cv = f_0 \tag{13_2}$$

or

$$v_{\alpha\alpha} - v_{\beta\beta} + Av_{\alpha} + Bv_{\beta} + Cv = f_0 \tag{13_3}$$

in hyperbolic case and

$$v_{\eta\eta} + Av_{\xi} + Bv_{\eta} + Cv = f_0 \tag{13_4}$$

in parabolic case.

To realize the possibility of reducing equation (8) to the canonical forms $(13_1), (13_2), (13_3)$ and (13_4) in all domain of its specification D (or, simply, "in large-scale") is quite difficult. However, the reasoning for realization of that possibility will be greatly simplified if we confine ourselves to a sufficiently small neighbourhood of a point (x, y) in domain D.

In fact, let us introduce the supposition that functions $\xi(x, y)$ and $\eta(x, y)$ have a Jacobian different from zero,

$$J = \frac{\partial(\xi, \eta)}{\partial(x, y)} .$$

The last requirement, as we know, guarantees local univalence of the mapping

$$\xi = \xi(x, y) , \quad \eta = \eta(x, y) .$$

As a result of the change of variables $\xi = \xi(x, y), \eta = \eta(x, y)$, Eq. (18) turns into

$$a_1 v_{\xi\xi} + 2b_1 v_{\xi\eta} + C_1 v_{\eta\eta} + d_1 v_3 + e_1 v_\eta + g_1 v = f_1 , \tag{8$_1$}$$

where

$$\begin{aligned} a_1(\xi, \eta) &= a\xi_x^2 + 2b\xi_x\xi_y + C\xi_y^2 , \\ b_1(\xi, \eta) &= a\xi_x\eta_x + b(\xi_x, \eta_y + \xi_y\eta_x) + C\xi_y\eta_y , \\ c_1(\xi, \eta) &= a\eta_x^2 + 2b\eta_x\eta_y + c\eta_y^2 , \\ v(\xi, \eta) &= u[x(\xi, \eta), y(\xi, \eta)] , \end{aligned} \tag{14}$$

and $x = x(\xi, \eta), y = y(\xi, \eta)$ is the inverse transformation of $\xi = \xi(x, y), \eta = \eta(x, y)$.

In the case that Eq. (8) is elliptic, i.e., if $b^2 - ac < 0$, for $\xi(x, y)$ and $\eta(x, y)$ we can take a solution with non-zero Jacobian J of the system of first order differential equations

$$\begin{aligned} a\xi_x + b\xi_y + \sqrt{ac - b^2}\eta_y &= 0 , \\ a\eta_x + b\eta_y - \sqrt{ac - b^2}\xi_y &= 0 . \end{aligned} \tag{15}$$

By virtue of (14) and (15) we have

$$a_1 = c_1 = \frac{ac - b^2}{a} (\xi_y^2 + \eta_y^2) \neq 0 , \quad b_1 = 0 .$$

Dividing all terms of Eq. (8_1) by the non-zero expression

$$\frac{ac - b^2}{a} (\xi_y^2 + \eta_y^2)$$

we obtain Eq. (13_1).

Notice that system (15) is equivalent to Eq. (19). This is easily verified as long as we apply the notation $\varphi = \xi + i\eta$, where i is the imaginary unit.

When $b^2 - ac > 0$, i.e., for hyperbolic equation (8), we can take the solutions of Eq. (9), which have a Jacobian J different from zero, for $\xi(x, y)$ and $\eta(x, y)$.

We assume that $a \neq 0$ (when $a = 0$ and $c = 0$ the reasoning is evidently to be changed). For this case we find from (9) and (14) that $a_1 = c_1 = 0, b_1 = 2\frac{ac-b^2}{a}\xi_y\eta_y \neq 0$ and, therefore, Eq. (8_1) has the form (13_2) after it is divided by $2b_1$. As a result of the new replacement $\alpha = \xi + \eta, \beta = \xi - \eta$, Eq. ($13_2$) becomes of the form ($13_3$).

What remains now is to consider the case of $b^2 - ac = 0$, i.e., when Eq. (8) is parabolic. The function $\xi(x, y)$ is to be taken different from the zero solution of Eq. (9). We select a function $\eta(x, y)$ so that the condition $a\eta_x^2 + 2b\eta_x\eta_y + c\eta_y^2 \neq 0$ be fulfilled. By virtue of $a\xi_x^2 + 2b\xi_x\xi_y + c\xi_y^2 = 0$ we have $a_1 = b_1 = 0$ from the first two equalities of (14). Thus, dividing Eq. (8_1) by $a\eta_x^2 + 2b\eta_x\eta_y + c\eta_y^2$ we obtain Eq. (13_4). Equation (9) is equivalent to two linear differential equations

$$a\varphi_x + \left(b + \sqrt{b^2 - ac}\right)\varphi_y = 0 ,$$
$$a\varphi_x + \left(b - \sqrt{b^2 - ac}\right)\varphi_y = 0$$

when $b^2 - ac > 0$, and to one equation

$$a\varphi_x + b\varphi_y = 0$$

when $b^2 - ac = 0$.

Therefore, we can assume that the functions $\xi(x, y)$ and $\eta(x, y)$ are solutions of the equations

$$a\xi_x + \left(b + \sqrt{b^2 - ac}\right)\xi_y = 0 , \quad a\eta_x + \left(b - \sqrt{b^2 - ac}\right)\eta_y = 0 \qquad (16_1)$$

if $b^2 - ac > 0$, and one of these functions, say, $\xi(x, y)$, is a solution of the equation

$$a\xi_x + b\xi_y = 0 \qquad (16_2)$$

if $b^2 - ac = 0$.

As it is well known, the problem of the existence of solutions of linear first order partial differential equations is extremely closely connected with the theory of ordinary first order differential equations. It has already been proved that under the condition that the functions a, b and c are sufficiently smooth, the system of linear partial differential equations (15) and the linear equations (16_1) and (16_2) have solutions of the form we need in some neighborhood of each point (x, y) of the domain D specified for Eq. (8). At the same time the possibility of reducing Eq. (8) to a canonical form $(13_1), (13_2), (13_3)$, or (13_4) in some neighborhood of a point (x, y) has also been proved.

2.3.2. Now let (8) be an equation of the mixed type in a domain D and its parabolic degeneration occurs along a curve σ. The equation of this curve can be written in the form

$$ac - b^2 = 0 \tag{17}$$

Suppose that a, b and c are all analytic functions of their arguments and σ is a simple analytic arc.

We write Eq. (17) in the form

$$H^n(x, y)G(x, y) - 0 \ , \tag{18}$$

where n is a natural number, $H(x, y) = 0$ is the equation of the curve σ along which the function $G(x, y) \neq 0$ and, in addition, H_x and H_y do not vanish simultaneously.

By σ_1 we denote the part of the arc σ along which either

$$aH_x^2 + 2bH_xH_y + cH_y^2 + cH_y^2 \neq 0 \tag{19}$$

or

$$aH_x^2 + 2bH_xH_y + cH_y^2 = 0 \ . \tag{20}$$

The fulfillment of the condition (19) means that the direction of the characteristic of the equation (8) at each point of σ_1 does not coincide with that of the tangent to the curve or that point, whereas the equality (20) means these two directions are coincident. If $\alpha(x, y)$ is the smallest angle between the tangent to the curve σ_1 at the point (x, y) and the characteristic direction outgoing from this point, then we suppose that either

$$\alpha(x, y) \neq 0 \tag{21}$$

for the whole of σ_1 or

$$\alpha(x, y) = 0 \tag{22}$$

at every point on it.

Let δ be a subdomain of the domain D, which contains the arc σ_1.

It can be proved that there is a transformation of independent variables, $\xi = \xi(x, y), \eta = \eta(x, y)$ which is nonsingular in domain δ, such that with its help Eq. (8) in this domain can be reduced to one of the following forms:

$$\eta^{2m+1} v_{\xi\xi} + v_{\eta\eta} + Av_\xi + Bv_\eta + cv = f_0 \tag{23}$$

or

$$v_{\xi\xi} + \eta^{2m+1} v_{\eta\eta} + Av_\xi + Bv_\eta + cv = f_0 \tag{24}$$

for odd $n = 2m + 1$, provided the condition (21) or (22) is fulfilled; and

$$\eta^{2m} v_{\xi\xi} \pm v_{\eta\eta} + Av_\xi + Bv_\eta + cv = f_0 \tag{25}$$

or

$$v_{\xi\xi} \pm \eta^{2m} v_{\eta\eta} + Av_\xi + Bv_\eta + cv = f_0 \tag{26}$$

for even $n = 2m$, provided the condition (21) or (22) is fulfilled.

In domain δ, Eqs. (23) and (24) are mixed type (elliptic-hyperbolic type) equations with parabolic degeneration when $\eta = 0$, whereas equations (25) and (26) are completely elliptic or completely hyperbolic for $\eta \neq 0$, but both degenerate parabolicly for $\eta = 0$. In a domain containing its own part of parabolicly degenerate line $\eta = 0$, it is impossible to reduce any one of the equations (23), (24), (25) and (26) to another one of them, or to an equation having the same form but another exponent of η, with the help of a nonsingular transformation of the variables.

§2.4 Systems of Linear Partial Differential Equations

2.4.1. As was already noted in §1 of this chapter, a system of linear second order partial differential equations where the number of equations, N coincides with the number of the unknown functions N_1, can be written in the form (4). In this case the characteristic form (2) has the form

$$K(\lambda_1, \cdots, \lambda_n) = \det \sum_{i,j=1}^{n} A_{ij}(x)\lambda_1\lambda_j \ . \tag{27}$$

Evidently, its degree is equal to $2N$.

For the system of form (4) the type division can be made in the same way in §1.

In the case of two independent variables, the representation (8) will be used for system (4) where a, b, c, d, e, g are square $N \times N$ matrices, and $u = (u_1, \cdots, u_N), f = (f_1, \cdots, f_N)$ are vectors. Without loss of generality we can suppose that $\det a \neq 0$. This can always be attained by a suitable change of the independent variables and unknown functions.

The ellipticity of system (8) means that the roots of the *characteristic polynomial* of degree 2N

$$\mathcal{P}(\lambda) = \det Q(\lambda) \ , \quad Q(\lambda) = a\lambda^2 + 2b\lambda + c \ ,$$

are all complex, while the hyperbolicity of this system means that all the roots of the polynomial $\mathcal{P}(\lambda)$ are real except for the exclusion of its parabolic degeneration at the same time.

The polynomial $\mathcal{P}(\lambda)$ can be written in the form

$$\mathcal{P}(\lambda) = \det \ a \prod_{i=1}^{l} (\lambda - \lambda_i)^{k_i} \ , \quad \sum_{i=1}^{l} k_i = 2N \ ,$$

when its root λ_i is of k_i multiplicity.

We prove that

$$k_1 \geq N - r_i \ , \quad i = 1, \cdots, l \ , \tag{28}$$

where

$$r_i = rank \ Q(\lambda_i) \ .$$

In fact, among the columns of matrix $Q(\lambda_i)$, namely $q_j(\lambda_i), j = 1, \cdots, N$, there are r_i columns being linearly independent because $r_i = rank \ Q(\lambda_i)$. Without loss of generality we assume that they are the $q_1, \cdots, q_{r_i}$-th columns. Then, we have

$$q_j(\lambda_i) = \sum_{p=1}^{r_i} c_{ijp} q_p(\lambda_1) \ , \quad j = r_{i+1}, \cdots, N \ .$$

Introducing the notation

$$q_{ij}^0(\lambda) \equiv q_j(\lambda) - \sum_{p=1}^{r_i} c_{ijp} q_p(\lambda) \ , \quad j = r_i + 1, \cdots, N \ ,$$

we have $q^0_{ij}(\lambda) = (\lambda - \lambda_i)q^1_{ij}(\lambda)$, where $q^1_{ij}(\lambda)$ is a polynomial vector of degree one in λ due to $q^0_{ij}(\lambda_j) = 0$.

In view of the fact that

$$\begin{aligned}\mathcal{P}(\lambda) &= \det\ Q(\lambda)\\ &= \det\left[q_1(\lambda), \cdots, q_{r_i}(\lambda), q^0_{ir_i+1}(\lambda), \cdots, q^0_{iN}(\lambda)\right]\\ &= (\lambda - \lambda_i)^{N-r_i} \det\left[q_1(\lambda), \cdots, q_{r_i}(\lambda), q^1_{ir_i+1}(\lambda), \cdots, q^1_{iN}(\lambda)\right] ,\end{aligned}$$

the validity of the stated assertion is obvious

By definition a hyperbolic system (28) is said to be *normally hyperbolic* if the inequality sign in it is excluded. A normally hyperbolic system is said to be *strictly hyperbolic* if the roots λ_i of polynomial $\mathcal{P}(\lambda)$ are all simple.

Evidently, it is not always possible to introduce a transformation of independent variables and unknown functions to reduce a system (8) to the forms of (13), (14), (15), (16), (23), (24), (25), (26), where A, B, C are $N \times N$ square matrices, and $u = (u_1, \cdots, u_N), f_0 = (f_1, \cdots, f_N)$ are vectors.

2.4.2. A system of linear first order partial differential equations

$$\sum_{i=1}^{n} A_i(x)\frac{\partial u}{\partial x_i} + B(x)u = F , \tag{29}$$

where A_i and B are $N \times N$ square matrices, $N > 1, u = (u_1, \cdots, u_N)$ is the required vector, and $F = (F_1, \cdots, F_N)$ is a given N-dimensional vector, is classified in accordance with the scheme stated in §2.1 of this chapter.

In the case of two independent variables $x_1 = x, x_2 = y$ writing the system (29) in the form

$$a\frac{\partial u}{\partial x} + b\frac{\partial u}{\partial y} + cu = F \tag{30}$$

and, without loss of generality, assuming $\det a \neq 0$, we consider the polynomial $P(\lambda)$ in λ of degree of N

$$\mathcal{P}(\lambda) = \det\ (a\lambda + b)$$

as above.

If $\mathcal{P}(\lambda)$ has no real roots at all system (30) is elliptic, and, if we exclude the case of parabolic degeneration, system (30) is hyperbolic when all roots of this polynomial are real. Moreover, the concepts of normal and strict hyperbolicity can be introduced this time.

The type classification for systems of linear partial differential equations of higher order can be dealt with similarly, and hence we shall not pay more attention to them here.

By a direct check, it is easy to be convinced that the system of equations (20) in shift component, which we had derived in §1.5 of chapter 1 for discussing the linear static theory of elasticity, is elliptic, and the so-called *generalized Cauchy-Riemann's system*

$$\begin{aligned}
\frac{\partial u}{\partial x} - \frac{\partial v}{\partial y} + a(x,y)u + b(x,y)v = F_1 \ , \\
\frac{\partial u}{\partial y} + \frac{\partial v}{\partial x} + c(x,y)u + d(x,y)v = F_2 \ ,
\end{aligned}$$

is also elliptic, while the system

$$\begin{aligned}
\frac{\partial u}{\partial x} - \frac{\partial v}{\partial y} + a(x,y)u + b(x,y)v = F_1 \ , \\
\frac{\partial u}{\partial y} - \frac{\partial v}{\partial x} + c(x,y)u + d(x,y)v = F_1 \ ,
\end{aligned}$$

is strictly hyperbolic.

2.4.3. Now we turn to the system of partial differential equations

$$q_1\frac{\partial q_1}{\partial x_1} + q_2\frac{\partial q_1}{\partial x_2} + \frac{1}{\rho}\frac{\partial p}{\partial x_1} = 0 \ , \quad q_1\frac{\partial q_2}{\partial x_1} + q_2\frac{\partial q_2}{\partial x_2} + \frac{1}{\rho}\frac{\partial p}{\partial x_2} = 0 \ , \tag{31}$$

$$\frac{\partial}{\partial x_1}(pq_1) + \frac{\partial}{\partial x_2}(\rho q_1) = 0 \ , \tag{32}$$

$$\frac{\partial q_2}{\partial x_1} - \frac{\partial q_1}{\partial x_2} = 0 \ , \tag{33}$$

which describes the planar stationary irrotational motion of a non-viscous compressible medium.

Taking into account (33) and the equation of state, $p = p(\rho)$, it follows from (31) that

$$\frac{1}{2}dg + \frac{1}{\rho}dp = \frac{1}{2}dg + \frac{1}{\rho}\frac{dp}{d\rho}d\rho = 0 \ , \tag{34}$$

where q is the following scalar function of the velocity q

$$q^2 = q_1^2 + q_2^2 \ .$$

Integrating (34) we obtain the equality

$$\frac{1}{2}q^2 + \int_{p_0}^{p} \frac{dp}{\rho} = \frac{1}{2}q^2 + \int_{\rho_0}^{\rho} \frac{1}{\rho}\frac{dp}{d\rho}d\rho = 0 \ , \quad p_0 = p(\rho_0) = \text{const} \ ,$$

which is well known as the so-called *Bernoulli's equation.*

Because the equation of state enters Bernoulli's equation, it follows that the density of the medium, ρ, is a function of q:

$$\rho = \rho(q) \ , \tag{35}$$

and, moreover, the quantity

$$c^2 = \frac{dp}{d\rho} = -\frac{pq}{p'(q)} \tag{36}$$

has the dimensions of velocity squared. The quantity c is called the *local velocity of sound*, the dimensionless quantity

$$M = \frac{q}{c} \tag{37}$$

is called *Mach number*, and the line on the plane of motion, i.e., the plane x_1 and x_2, along which $M = 1$ is called a *sound line.* the value of q for which $M = 1$ is denoted by q_k.

In discussion we introduce the *potential of velocity* $\varphi(x_1, x_2)$ and the *stream function* $\psi(x_1, x_2)$ with the help of the equalities

$$q_1 = \frac{\partial \varphi}{\partial x_1} = \frac{\rho_0}{\rho}\frac{\partial \psi}{\partial x_2} \ , \quad q_2 = \frac{\partial \varphi}{\partial x_2} = -\frac{\rho_0}{\rho}\frac{\partial \psi}{\partial x_1} \ , \tag{38}$$

which clearly fulfill the equations (32) and (33), q^2 then has the form.

$$q^2 = (grad\ \varphi)^2 = \left(\frac{\rho_0}{\rho}\right)^2 (grad\ \psi)^2 . \tag{39}$$

2.4.4. The plane of variables x_1, x_2, on which we study the motion (flow) of a compressible medium is called a *physical plane.* Because the velocity vector (q_1, q_2) of a particle which has coordinates (x_1, x_2), corresponding to every point (x_1, x_2) in the medium, has been placed, learning the law of the motion means that the functions $q_1(x_1, x_2)$ and $q_2(x_1, x_2)$ are known, or, equivalently, the functions $q(x_1, x_2)$ and $\theta(x, x_2)$ defined by the equalities

$$q_1 = q \cos\ \theta\ , \quad q_2 = q \sin\ \theta \tag{40}$$

are known. The plane where the variables q, θ take values is usually called the the *hodograph plane.*

By virtue of (39) the equality (38) is a system of nonlinear partial differential equations with respect to φ and ψ. We now prove that this system becomes linear as a result of a transition from the physical plane to a hodograph plane.

Indeed, by virtue of (38) and (40) we have

$$\begin{aligned} dx_1 + idx_2 &= e^{i\theta} \left(e^{-i\theta} dx_1 + ie^{-i\theta} dx_2\right) \\ &= \frac{e^{i\theta}}{q} \left(d\varphi + i\frac{\rho_0}{\rho} d\psi\right) , \end{aligned}$$

or equivalently,

$$dx_1 + idx_2 = \frac{e^{i\theta}}{q} \left[\left(\frac{\partial\varphi}{\partial q} + i\frac{\rho_0}{\rho}\frac{\partial\psi}{\partial q}\right) dq + \left(\frac{\partial\varphi}{\partial\theta} + i\frac{\rho_0}{\rho}\frac{\partial\psi}{\partial\theta}\right) d\theta\right] . \tag{41}$$

The necessary and sufficient conditions for the expression (41) to be a total differential are

$$\frac{\partial\varphi}{\partial\theta} - \frac{q\rho_0}{\rho}\frac{\partial\psi}{\partial q} = 0 \tag{42}$$

and

$$\frac{\partial\psi}{\partial q} + \frac{\rho_0}{\rho q}\left(1 + \frac{\rho' q}{\rho}\right)\frac{\partial\psi}{\partial\theta} = 0 . \tag{43}$$

Because of

$$\frac{\rho' q}{\rho} = -M^2 ,$$

on account of (36) and (37) the equality (43) is equivalent to

$$\frac{\partial \psi}{\partial q} + \frac{\rho_0}{\rho} \frac{1 - M^2}{q} \frac{\partial \psi}{\partial \theta} = 0 . \tag{44}$$

The equalities (42) and (44) in φ and ψ are a linear system of first order partial differential equations because the variables ρ and M are functions of q only on account of (35), (36) and (37).

Instead of q, we can introduce a new dimensionless independent variable v defined by

$$v = -\int_{q_k}^{q} \frac{\rho(\tau)}{\rho_0 \tau} d\tau .$$

and consider

$$\frac{\partial}{\partial q} = -\frac{\rho}{\rho_0 q} \frac{\partial}{\partial v} .$$

The system (42)–(44) can then be written in the form

$$\frac{\partial u_1}{\partial \theta} + \frac{\partial u_2}{\partial v} = 0 , \quad \frac{\partial u_1}{\partial v} - k(v) \frac{\partial u_2}{\partial \theta} = 0 , \tag{45}$$

with $u_1 \varphi(q, \theta), u_2 = \psi(q, \theta)$ and

$$k(v) = \frac{\rho_0^2}{\rho^2}(1 - M^2) .$$

The domains in the hydograph plane where $k > 0$ ($k < 0$) are called the domains of *subsonic motion* the domains of (*supersonic motion*), and the domain whose intersection with the subsonic line is not empty is called the domain of *transonic motion*. In accordance with the classification given before, the system (45) is elliptic in a domain of *subsonic motion*, hyperbolic in a domain of *supersonic motion*, and is related to a system of mixed type equations in the domain of *transonic motion*.

Eliminating the function u_1 from system (45) gives the *Chaplygin equation*

$$k(v)\frac{\partial^2 u_2}{\partial \theta^2} + \frac{\partial^2 u_2}{\partial v^2} = 0 .$$

§2.5 Concept of Well-posedness in Formulating Problems of Partial Differential Equations

2.5.1. There are partial differential equations for which the set of solutions are very narrow or even empty. For example, the set of real solutions for the equation

$$\sum_{i=1}^{n} \left(\frac{\partial u}{\partial x_i}\right)^2 + 1 = 0$$

has no real solutions at all. Partial differential equations occurring in applications, however, usually have an extensive family of solutions.

In the theory of ordinary differential equations there is a concept of general solution. The essence of this concept is illustrated most vividly by the example of the linear equation of the n-th order.

$$\frac{d^n y}{dx^n} + \sum_{k=1}^{n} a_k(x)\frac{d^{n-k} y}{dx^{n-k}} = 0 , \tag{46}$$

where $a_k(x), k = 1, \cdots, n$, are continuous functions given on some interval $x_0 \leq x \leq x_1$. As we know, this equation has a set of n linear independent solutions $y_1(x), \cdots y_n(x)$ with the help of which any one of its solutions, $y(x)$, can be expressed in the form

$$y(x) = \sum_{k=1}^{n} c_k y_k(x) , \tag{47}$$

where $c_k, k = 1, \cdots, n$, are constants. It is in this sence that the expression (47) where $c_1, \cdots, c_n$ are arbitrary constants is a general solution of Eq. (46).

Although in the theory of partial differential equations there is no similar criterion for generality of a solution, it is still possible in some cases to write down a formula which plays the role of general solution. For example, the oscillating string equation

$$\frac{\partial^2 u}{\partial t^2} - \frac{\partial^2 u}{\partial x^2} = 0 \tag{48}$$

can be rewritten in the form

$$\frac{\partial^2 v}{\partial\xi\partial\eta} = 0$$

by replacement of the variables

$$\xi = x + t, \eta = x - t, v(\xi, \eta) = u\left(\frac{\xi + \eta}{2}, \frac{\xi - \eta}{2}\right) .$$

By integrating we obtain from it

$$v(\xi, \eta) = f_1(\xi) + f_2(\eta) .$$

Therefore, the formula

$$u(x, t) = f_1(x + t) + f_2(x - t) \tag{49}$$

gives a regular solution of Eq. (48) for arbitrary functions f_1, f_2 which are twice continuously differentiable.

It can be verified by direct checking that the expression

$$u(x, y) = \varphi(z) + \bar{\varphi}(\bar{z}) , \tag{50}$$

where $\varphi(z)$ is an arbitrary analytic function of a complex variable $z = x + iy$, represents a regular solution of Laplace's equation

$$\frac{\partial^2 u}{\partial x^2} + \frac{\partial^2 u}{\partial y^2} = 0 . \tag{51}$$

An arbitrary solution of the equations (48) and (51) can be expressed respectively in the forms (49) and (50). Thus, these formulae play the role of general solutions for the above-mentioned equations.

2.5.2. In some problems which are modelled in terms of partial differential equations there are some additional requirements for the solutions of the given equation. That is, for them the problems posed are associated with certain conditions on those manifolds whose dimensionality is less than that of domain D of specification for the equation by one, and, furthermore, the manifolds, called carriers of given data, are required to be either in the domain D or on its boundary ∂D. Whenever a solution which is subject to these conditions exists uniquely and is stable in a definite sense (we will talk about this below), we say that this problem is *well-posed* or *correctly formulated*. It is very important to note that the well-posedness for formulating problems of partial differential equations essentially depends on the type of the equation considered.

We now turn our attention to some examples of well-posed problems for the simplest hyperbolic, elliptic and parabolic types of equations.

§2.6 Wave Equation

2.6.1. As already established in §1.3 of Chapter 1, in the linear oscillation theory of a membrane with fixed edges, its shift $u(x, y, t)$ at point (x, y) is a solution of the equation

$$\frac{\partial^2 u}{\partial t^2} - \frac{\partial^2 u}{\partial x^2} - \frac{\partial^2 u}{\partial y^2} = 0 \,, \tag{52}$$

which is an equation of hyperbolic type according to the definition given in §2.2.

By direct measurement an observer is able to compute for the initial instant t_0 of time t the function $u(x, y, t_0)$—the definite initial position of the membrane, and also the initial velocity $\frac{\partial u(x,y,t)}{\partial t}|_{t=t_0} = \psi(x, y)$ of its motion.

Therefore, the function $u(x, y, t)$ describing the oscillation of the membrane, which is fixed along its edges, in a domain G in the plane of the variables x, y should be a regular solution of Eq. (52), satisfying the conditions

$$u(x, y, t_0) = \varphi(x, y) \,, \quad \left.\frac{\partial u(x, y, t)}{\partial t}\right|_{t=t_c} = \psi(x, y) \,. \tag{53}$$

The equation (52), with $n = 2, x_1 = x, x_2 = y$ is a special case of the wave equation with n spatial variables $x_1, \ldots, x_n$:

$$\frac{\partial^2 u}{\partial t^2} - \Delta u = 0 \,, \quad \Delta = \sum_{i=1}^{n} \frac{\partial^2}{\partial x_i^2} \,. \tag{54}$$

The solutions of equation (54) are usually known as *waves*.

When $n = 1$ or $n = 3$ equation (54) models oscillations of, respectively, 1-dimensional continuum (for instance, a string) and 3-dimensional continuum in orthogonal Cartesian coordinates. Equation (54) is hyperbolic for all values of the independent variables.

Let G be a domain of points $x = (x_1, \dots, x_n)$ of n-dimensional space E_n, which lies in the hyperplane $t = t_0$ in the space E_{n+1} of points (x, t), and $\varphi(x)$ and $\psi(x)$ be real functions given on G.

The problem to determine the regular solution $u(x,t)$ of equation (54) with given

$$u(x, t_0) = \varphi(x)\ , \quad \left.\frac{\partial u(x,t)}{\partial t}\right|_{t=t_0} = \psi(x)\ , \quad x \in G\ , \tag{55}$$

is known as the *Cauchy Problem*, and the conditions formulated in the form of equality (55) are called the *initial conditions*.
In such a formulation for the Cauchy problem the domain G in the hyperplane $t = t_0$ of space E_{n+1} is the data carrier.

2.6.2. A solution of problem (54)–(55) is constructed by quadratures. Here we confine ourselves to considering the cases of $n = 3, 2, 1$ and suppose at first that $n = 3$.

Now we prove that the function $u_0(x,t)$ defined by the formula

$$u_0(t,t) = \int_S \frac{\mu(y)}{|y - x|} ds_y\ ,$$

where $|y - x|$ is the distance between points $x = (x_1, x_2, x_3)$ and $y(y_1, y_2, y_3)$, S is the sphere $|y - x|^2 = t^2$, and μ is a given real function defined on some range of variables y_1, y_2, y_3 having continuous derivatives of the second order, is a regular solution of the equation

$$\frac{\partial^2 u}{\partial t^2} - \sum_{i=1}^{3} \frac{\partial^2 u}{\partial x_i^2} = 0\ . \tag{56}$$

By the replacement of variables $y_i - x_i = t\xi_i, i = 1, 2, 3$, we rewrite the expression of $u_0(x,t)$ in the form

$$u_0(x,t) = t \int_\sigma \mu(x_1 + t\xi_1, x_2 + t\xi_2, x_3 + t\xi_3) d\sigma_\xi\ , \tag{57}$$

where σ is the unit ball $|\xi| = 1$ and

$$d\sigma_\xi = \frac{ds_y}{t^2} = \frac{ds_y}{|y - x|^2} \tag{58}$$

is its element of area.

The function $u_0(x,t)$ represented by the formula (57) obviously has continuous second derivatives and, moreover,

$$\sum_{i=1}^{3} \frac{\partial^2 u_0}{\partial x^2} = t \int_\sigma \sum_{i=1}^{3} \frac{\partial^2 \mu}{\partial y_i^2} d\sigma_\xi \ . \tag{59}$$

Besides,

$$\begin{aligned} \frac{\partial u_0}{\partial t} &= \int_\sigma \mu(x_1 + t\xi_1, x_2 + t\xi_2, x_3 + t\xi_3) d\sigma_3 + t \int_\sigma \sum_{i=1}^{3} \frac{\partial \mu}{\partial y_i} \xi_i d\sigma_3 \\ &= \frac{1}{t} u_0 + \frac{1}{t} I \ , \end{aligned} \tag{60}$$

where

$$I = \int_S \left[\frac{\partial \mu}{\partial y_1} \nu_1 + \frac{\partial \mu}{\partial y_2} \nu_2 + \frac{\partial \mu}{\partial y_3} \nu_3 \right] ds_y \tag{61}$$

and $\nu(y)$ is the unit outward normal vector to S at the point y.

Differentiating the equality (60) in t, we obtain

$$\begin{aligned} \frac{\partial^2 u_0}{\partial t^2} &= -\frac{1}{t} u_0 + \frac{1}{t} \frac{\partial u_0}{\partial t} - \frac{1}{t^2} I + \frac{1}{t} \frac{\partial I}{\partial t} \\ &= -\frac{1}{t^2} u_0 + \frac{1}{t} \left(\frac{1}{t} u_0 + \frac{1}{t} I \right) - \frac{1}{t^2} I + \frac{1}{t} \frac{\partial I}{\partial t} \\ &= \frac{1}{t} \frac{\partial I}{\partial t} \ . \end{aligned} \tag{62}$$

Applying the Gauss-Ostrogradskii formula we rewrite the formula (61) in the form

$$I = \int \sum_{i=1}^{3} \frac{\partial^2 \mu}{\partial y_i^2} d\tau_y \ , \tag{63}$$

where the integration is taken over the ball $|y - x|^2 < t^2$.

Turning the Cartesian coordinates y_1, y_2, y_3 into spherical coordinates ρ, θ, φ, we have from (63)

$$I = \int_0^t d\rho \int_0^\pi d\theta \int_0^{2\pi} \Delta\mu\rho^2 \sin\theta d\varphi \ , \tag{64}$$

where $\rho^2 \sin\theta d\rho d\theta d\varphi = d\tau_y$.

On account of $\sin\theta d\theta d\varphi = d\sigma_\xi$, we find from (64) that

$$\begin{aligned} \frac{\partial I}{\partial t} &= t^2 \int_0^\pi d\theta \int_0^{2\pi} \Delta\mu \sin\theta d\varphi \\ &= t^2 \int_\sigma \sum_{i=1}^3 \frac{\partial^2 \mu}{\partial y_i^2} d\sigma_\xi \ . \end{aligned} \tag{65}$$

By virtue of (65) and (62) we have

$$\frac{\partial^2 u_0}{\partial t^2} = t \int_\sigma \sum_{i=1}^3 \frac{\partial^2 \mu}{\partial y_i^2} d\sigma_\xi \ . \tag{66}$$

On the basis of (59) and (66) we conclude that $u_0(x,t)$ is a regular solution fo Eq. (54) when $n = 3$.

From the argument adduced above it is clear that this conclusion is true regardless of the sign of the variable t.

Introduce the notation

$$tM(\mu) = \frac{1}{4\pi} u_0(x,t) \ , \tag{67}$$

where

$$M(\mu) = \frac{1}{4\pi} \int_\sigma \mu(x_1 + t\xi_1, x_2 + t\xi_2, x_3 + t\xi_3) d\sigma_\xi \tag{68}$$

due to (57).

Taking into account the formula (58), the formula (68) from $M(\mu)$ can be written in the form

$$M(\mu) = \frac{1}{4\pi t^2} \int_s \mu(y) ds_y \ . \tag{69}$$

Because $4\pi t^2$ is the area of the sphere $S : |y - x|^2 = t^2$, the representation (69) means that $M(\mu)$ represents the integral average of the function μ over the sphere S.

It is clear that, along with $tM(\mu)$, the function $\frac{\partial}{\partial t}[tM(\mu)]$ is also a regular solution of equation (56) under the condition that μ has continuous derivatives of the third order.

Therefore, the function

$$u(x,t) = tM(\psi) + \frac{\partial}{\partial t}[tM(\varphi)] \tag{70}$$

is a regular solution of equation (56) if $\psi(x)$ and $\varphi(x)$ have continuous derivatives of the second order and the third order, respectively.

It is easy to see that the function $u(x,t)$ defined by the formula (70) also satisfies the initial conditions (55) at $t_0 = 0$. In fact, from (68) and (70) we obtain

$$u(x,0) = \frac{1}{4\pi}\int_\sigma \varphi(x)d\sigma_\xi = \varphi(x) \ ,$$

$$\left.\frac{\partial u(x,t)}{\partial t}\right|_{t=0} = \frac{1}{4\pi}\int_\sigma \psi(x)d\sigma_\xi = \psi(x) \ .$$

When the initial conditions are taken at $t = t_0$ in the form (55), the function $u(x,t-t_0)$ is clearly a solution of the problem (55)–(56), where $u(x,t)$ is given by (70).

The expression (70) is known as *Kirchhoff's formula.*

By the use of Kirchhoff's formula (70) a solution $u(x,t)$ of the Cauchy problem (55)–(56) is determined at those points in the space E_4 of variables x_1, x_2, x_3 and t that have such a property: the intersection of the cone K : $|y-x|^2-(\tau-t)^2 = 0$, which has a vertex at the point (x,t), and the hyperplane $\tau = 0$ belong to the domain G. When the carrier G given by (55) coincides with the hyperplane $t = 0$ in the space E_4, Kirchhoff's formula then gives a solution of the Cauchy problem (55)–(56) at all finite points (x,t) of the space E_4.

Because of

$$\frac{\partial}{\partial t}[tM(\varphi)] = M(\varphi) + \frac{1}{4\pi t}\int_S \frac{\partial\varphi}{\partial\nu_y}ds_y \ ,$$

by taking into account (67) and (68), it follows from Kirchhoff's formula (70) that, in order to obtain a solution $u(x,t)$ of the problem (55)–(56) at a point (x,t), it is enough to know only the values of φ, $\frac{\partial\varphi}{\partial\nu}$ and ψ, on the sphere

$|y - x|^2 = t^2$. This fact is known as the *Huygen's principle* in the theory of sound.

2.6.3. When $n = 2$, a solution $u(x,t)$ of the problem (54)–(55) can be obtained from Kirchhoff's formula by the descent method. The essence of this method is that, when the functions φ and ψ on the right-hand side of the formula (70) depend on two variables x_1 and x_2, this formula gives

$$\begin{aligned} u(x,t) &= U(x_1, x_2, t) \\ &= \frac{1}{4\pi t}\int_{|y|^2=t^2} \psi(x_1+y_1, x_2+y_2)ds_y \\ &+ \frac{1}{4\pi t}\frac{\partial}{\partial t}\left[\frac{1}{t}\int_{|y|^2=t^2} \varphi(x_1+y_1, x_2+y_2)ds_y\right], \end{aligned} \tag{71}$$

which is independent of x_3, satisfying the equation (54) in x_1, x_2 and t as well as the initial conditions (55).

The projection dy_1dy_2 of the element of area ds_y of the sphere $|y|^2 = t^2$ on the disk $y_1^2 + y_2^2 < t$ is expressed in terms of ds_y by the formula

$$dy_1dy_2 = ds_y \;\cos\, \widehat{i_3, \nu} = \frac{y_3}{t}ds_y ,$$

where i_3 is the unit vector of the axes x_3, and ν is the normal vector of the sphere $|y|^2 = t^2$ at the point (y_1, y_2, y_3). Therefore, taking into account the circumstance that for computing the integrals on the right-hand side of formula (71) we must project the upper one-half (i.e. $y_3 < 0$) as well as the lower one-half of the sphere $|y|^2 = t^2$ on the disk $y_1^2 + y_2^2 < t$, this formula will be written in the form

$$\begin{aligned} u(x_1, x_2, t) &= \frac{1}{2\pi}\int_d \frac{\psi(y_1, y_2)dy_1dy_2}{\sqrt{t^2 - (y_1 - x_1)^2 - (y_2 - x_2)^2}} \\ &+ \frac{1}{2\pi}\frac{\partial}{\partial t}\int_d \frac{\varphi(y_1, y_2)dy_1dy_2}{\sqrt{t^2 - (y_1 - x_2)^2 - (y_2 - x_2)^2}}, \end{aligned} \tag{72}$$

where d is the disk $(y_1 - x_2)^2 + (y_2 - x_2)^2 < t^2$.

The formula (72) which gives a solution of the Cauchy problem (54)–(55) with $n = 2$ is called *Poisson's formula*. From this formula it can be seen that for the case of $n = 2$ merely knowing the values of $\varphi(x_1, x_2)$ and $\psi(x_1, x_2)$ at the circumference $(y_1 - x_2)^2 + (y_2 - x_2)^2 = t^2$ is not enough for determining a solution $u(x,t)$ of the Cauchy problem (54)–(55) at a point (x,t). In the definition of $u(x_1,x_2,t)$ at a point (x_1,x_2,t), the values initially given at all points of the disk d are involved. This indicates that in the case $n = 2$, Huygen's principle does not hold in wave processes.

2.6.4. If $n = 1$, the formula that gives a solution of the problem (54)–(55) can clearly be obtained from Poisson's formula (72) by the use of the descent method again. But, to achieve this purpose we need to use the formula (49) which represents the solutions of Eq. (48). By substituting the expression (49) for the function $u(x,t)$ in the initial conditions (55), we have

$$f_1(x) + f_2(x) = \varphi(x) \ , \quad f_1'(x) - f_2'(x) = \psi(x) \ , \quad x \in G \ .$$

By integrating the latter of those just obtained, we find

$$f_1(x) - f_2(x) = \int_{x_0}^{x} \psi(y)dy \ .$$

Therefore, for f_1, and f_2 we have expressions

$$f_1(x) = \frac{1}{2}\left[\varphi(x) + \int_{x_0}^{x} \psi(y)dy\right] \ , \quad f_2(x) = \frac{1}{2}\left[\varphi(x) + \int_{x}^{x_0} \psi(y)dy\right] \ ,$$

and after introducing them into right-hand side of (49) we obtain a solution of the problem (54)–(55) with $n = 1$:

$$u(x,t) = \frac{1}{2}\left[\varphi(x+t) + \varphi(x-t) + \int_{x-t}^{x+t} \psi(y)dy\right] \ , \quad x_1 = x \ . \tag{73}$$

The formula (73) is called *d'Alembert's formula.*

2.6.5. For the given data carrier we take the plane $t = \tau_1$, instead of $t = 0$, and by $v(x,t,\tau), x = (x_1,x_2,x_3)$ we denote a solution of the wave equation (54) that satisfies the initial conditions

$$v(x,\tau_1,\tau_1) = 0 \ , \quad v_t(x,t,\tau_1)|_{t=\tau_1} = g(x,\tau_1) \ ,$$

where $g(x,\tau_1)$ is a given real function having continuous partial derivatives of the second order.

By substituting $t - \tau_1$ for t from Kirchhoff's formula (70) we have

$$v(x, t, \tau_1) = \frac{1}{4\pi(t - \tau_1)} \int_{|y-x|^2=(t-\tau_1)^2} g(y, \tau_1) ds_y .$$

It is not difficult to see that the function

$$u(x, t) = \int_0^t v(x, t, \tau_1) d\tau_1 \tag{74}$$

is a solution of the Cauchy problem

$$u(x, 0) = 0 , \quad u_t(x, t)|_{t=0} = 0$$

for the *non homogeneous wave equation*

$$u_{tt} - \Delta u = g(x, t) .$$

As a result of the replacement of the integral argument $t = t - \tau_1$, the formula (74) is rewritten in the form

$$\begin{aligned} u(x, t) &= \frac{1}{4\pi} \int_0^t d\tau \int_{r^2=\tau^2} g(y, t - \tau) \frac{ds_y}{\tau} \\ &= \frac{1}{4\pi} \int_{r^2<t^2} \frac{g(y, t - r)}{r} d\tau_y , \end{aligned} \tag{75$_1$}$$

where $r = |y - x|$.

The function $u(x, t)$ defined by the formula (75_1) coincides with the potential of volume mass distributed on the ball $r^2 < t^2$ with a density $g(y, t - r)$. In view of the fact that the volume of function g to be used in the formula (75_1) is that at the time $t - \tau$ which lags behind the instant of observation on the wave, this potential is called a *retared potential.*

In the case of two spatial variables, we can repeat the conclusion stated before when we derived the formula (74) so that the formula

$$u(x_1, x_2, t) = \frac{1}{2\pi} \int_0^t d\tau \int_d \frac{g(y_1, y_2, \tau) dy_1 dy_2}{\sqrt{(t - \tau)^2 - (y_1 - x_1)^2 - (y_2 - x_2)^2}} , \tag{75$_2$}$$

where d is the circle $(y_1 - x_1)^2 + (y_2 - x_2)^2 < (t - \tau)^2$, gives the solution of the homogenous Cauchy problem

$$u(x_1, x_2, 0) = 0 , \quad u_t(x_1, x_2, t)|_{t=0} = 0$$

for the nonhomogeneous wave equation with two space variables

$$u_{tt} - u_{x_1x_2} - u_{x_2x_2} = g(x_1, x_2, t) .$$

Similarly, it can be verified that the function

$$u(x,t) = \frac{1}{2}\int_0^t d\tau \int_{x-t+\tau}^{x+t-\tau} g(\tau_1, \tau) d\tau_1 \qquad (75_3)$$

is the solution of the nonhomogeneous equation for an oscillation string,

$$u_{tt} - u_{xx} = g(x,t) ,$$

satisfying the initial conditions

$$u(x,0) = 0 , \quad u_t(x,t)|_{t=0} = 0 .$$

The functions $g(x_1, x_2, t)$ and $g(x,t)$ in (75_2) and (75_3) are assumed to have continuous partial derivatives of the second and first orders respectively.

2.6.6. For a point (x,t) in the space E_{n+1}, a set of points in the domain G where the value of $u(x,t)$ of the problem (54)–(55) is completely determined by the initial values (55) on it is called the ***domain of dependence*** of the point (x,t). Of course, the points where the values of the functions $\varphi(x)$ and $\psi(x)$ are not involved in determining $u(x,t)$ at the point (x,t) do not belong to the domain of dependence of this point.

On the basis of the formulae (70), (72) and (73) we conclude that for $n = 3$ the dependence domain is determined in accordance with Huygens's principle, and for $n = 2$ and $n = 1$, the dependence domains of a point (x,t) are respectively the circle $|y-x|^2 \leq t^2$ and the segment $|y-x|^2 \leq t^2$ belonging to G.

The values of $\varphi(x)$ and $\psi(x)$ on G influence the values of $u(x,t)$ at all those points $(x,t) \in E_{n+1}$ having the property that the intersection of G and $\{|y-x|^2 < t^2\}$ is not empty. The set of those points is usually named the ***domain of influence.***

The set of those points $(x,t) \in E_{n+1}$ for which the solution $u(x,t)$ of the problem (54)–(55) is completely determined by the values of $\varphi(x)$ and $\psi(x)$ on G is called the ***domain of definition*** of $u(x,t)$. Again, from the formulae (70),

(72) and (73), the domain of definition of the solution $u(x,t)$ consists only of those points $(x,t) \in E_{n+1}$: where (1) for $n = 3$ the sphere $|y-x|^2 = t^2$, i.e., the intersection of a cone with vertex at the point (x,t),

$$K: \quad |y-x|^2 = (\tau - t)^2 , \tag{76}$$

and the hyperplane $\tau = 0$ belongs to G: (2) for $n = 2$ not only the circumference $|y-x|^2 = t^2$, i.e., the intersection of the cone K and the plane $\tau = 0$, but the whole disk $|y-x|^2 \leq t^2$ belongs to G; finally (3) for $n = 1$ not only the points $(x-t, 0)$ and $(x+t, 0)$, i.e., the intersection of the straight lines (degenerate cones) $y - x = \tau - t$ and $y - x = t - \tau$ passing through the point (x,t) and the straight line $\tau = 0$, but also the whole rectilinear segment between the two points, belongs to G.

2.6.7. Now we will prove that in its own domain of definition the solution of problem (54)–(55) is unique. Assuming that there exist two solutions $u_1(x,t)$ and $u_2(x,t)$ for this problem, by virtue of the linear equation (54) and the initial conditions (55) the function $u(x,t) = u_1(x,t) - u_2(x,t)$ is a solution of Eq. (54) and satisfies the homogeneous initial conditions

$$u(x,0) = 0 , \quad \frac{\partial u(x,t)}{\partial t}|_{t=0} = 0 . \tag{77}$$

Therefore, the uniqueness of the solutions for the problem (54)–(55) will be proved if we can show that the problem (54)–(77) has only the trivial (i.e. identically zero) regular solution.

Let D be a finite subdomain of points in the domain of definition of the regular solution $u(x,t)$ for the problem (54)–(77), and be bounded by the cone K and the hyperplane $\tau = 0$.

It is clear that for any regular solution $u(y,\tau)$ of Eq. (54), in which $t = \tau, x = y$ the identity

$$\begin{aligned} &-2\frac{\partial u}{\partial \tau}\left(\sum_{i=1}^{n} \frac{\partial^2 u}{\partial y_i^2} - \frac{\partial^2 u}{\partial \tau^2}\right) \\ &= -2\sum_{i=1}^{n} \frac{\partial}{\partial y_i}\left(\frac{\partial u}{\partial y_1}\frac{\partial u}{\partial \tau}\right) + \frac{\partial}{\partial \tau}\left[\sum_{i=1}^{n}\left(\frac{\partial u}{\partial y_i}\right)^2 + \left(\frac{\partial u}{\partial \tau}\right)^2\right] \\ &= 0 \end{aligned}$$

holds. Integrating this identity over D and applying the Gauss-Ostrogradskii formula we find that

$$\int_{\partial D}\left[-2\sum_{i=1}^{n}\frac{\partial u}{\partial\tau}\frac{\partial u}{\partial y_i}y_{i\nu}+\left(\frac{\partial u}{\partial\tau}\right)^2\tau_\nu+\sum_{i=1}^{n}\left(\frac{\partial u}{\partial y_i}\right)^2\tau_\nu\right]ds=0\,, \tag{78}$$

where $y_{i\nu}, i=1,\dots,n$, and τ_ν is the cosine of the unit normal vector ν to ∂D at point (y,τ).

By virtue of the conditions (77), the integral on the left-hand side of (78) taken over that part of ∂D where $\tau=0$, is equal to zero. Over the remaining part M of the boundary of domain D, we rewrite the equality (78) in the form

$$\int_M\frac{1}{\tau_\nu}\sum_{i=1}^{n}\left(\frac{\partial u}{\partial y_i}\tau_\nu-\frac{\partial u}{\partial\tau}y_{i\nu}\right)^2ds=0\,. \tag{79}$$

Because of $\tau_\nu=$ constant, we obtain from (79)

$$\frac{\partial u}{\partial y_i}\tau_\nu-\frac{\partial u}{\partial\tau}y_{i\nu}=0\,,\quad i=1,\dots,n\,. \tag{80}$$

The fulfillment of the equalities (80) means that on M the instrinsic derivatives of $u(y,\tau)$ along n linearly independent directions are equal to zero. Hence $u(y,\tau)=$ constant on M. From this, it thus follows that $u(y,\tau)=0$ on the whole ∂D due to $u(y,0)=0$; in particular, $u(x,t)=0$. In view of the fact that the (x,t) is an arbitrary point of the domain of definition for the solution $u(x,t)$ of the problem (54)–(77), it is equal to zero identically.

From the formulae (70), (72) and (73) it follows immediately that an infinitesimal alteration of the initial values (55) implies an infinitesimal alteration of the solution for the problem (54)–(55). In that sense the solution of the problem (54)–(55) is stable.

Thus, we come to the conclusion that the Cauchy problem with initial conditions (55) of the wave equation (54) is well-posed.

It would be incorrect to say that the problem (54)–(55) is a boundary problem since the domain of definition of its solution contains its own carrier G inside.

When the carrier G given by (55) coincides with the hyperplane $t=0$, the domain of definition of solution for the problem (54)–(55) coincides with the set of all finite points in the space E_{n+1}.

There are formulae which give by quadratures the solution of the problem (54)–(55) for both arbitrary even and odd natural numbers n. They are known as *Matison's formulae.*

§2.7 Laplace Equation

2.7.1. In Chapter 1 we saw that Laplace's equation (38) written in orthogonal Cartesian coordinates,

$$\sum_{i=1}^{n} \frac{\partial^2 u}{\partial x_i^2} = 0 , \tag{81}$$

is elliptic in all the space E_n of points $x = (x_1, \dots, x_n)$. It is clear that it is uniformly elliptic because the quadratic form (5) corresponding to this equation has the canonical form

$$Q(\lambda_1, \dots, \lambda_n) = \sum_{i=1}^{n} \lambda_i^2 .$$

A regular solution of equation (81) in a domain D of the space E_n, $u(x) = u(x_1, \dots, x_n)$, is called a *harmonic function.*

It follows from the linearity of Laplace's equation that harmonic functions have the following simplest properties: (a) if $u_1(x), k = 1, \dots, m$, are harmonic then the function $u(x) = \sum_{k=1}^{m} C_k u_k(x), c_k =$ constant is also harmonic; (b) along with a function $u(x)$ which is harmonic in a domain D, the function $u(\lambda C x + h)$, where λ is a scalar real constant, C is a constant real orthogonal $n \times n$ matrix and $h = (h_1, \dots, h_n)$ is a constant real vector, with the assumption that $\lambda C x + h$ are in the domain D for all points x, is also a harmonic function; (c) if a function $u(x)$ is harmonic in a domain D, then the function

$$v(x) = |x|^{2-n} u\left(\frac{x}{|x|^2}\right) , \quad x^2 = |x|^2 = \sum_{i=1}^{n} x_i^2 ,$$

is also harmonic at all points where it is defined.

When the domain D contains an infinite point, the definition of a harmonic function mentioned above must be modified as the concept of derivative at an infinite point makes no sense.

We say that a function $u(x)$ is harmonic in a neighbourhood of an infinite point (i.e., outside a closed ball with a sufficiently large radius $R : |x| \leq R$), if the function

$$v(y) = |y|^{2-n} u\left(\frac{y}{|y|^2}\right) , \quad y = \frac{x}{|x|^2} , \tag{82}$$

which is defined at the point $y = 0$ by the extension $\lim\limits_{y \to 0} v(y)$, is harmonic in a neighbourhood of the point $y = 0$.

As a result of the change $y = \frac{x}{|x|^2}$ we obtain from (82)

$$u(x) = |x|^{2-n} v\left(\frac{x}{|x|^2}\right) .$$

In accordance with this, by saying that the solution of Eq. (81) is *regular in a neighbourhood of an infinite point* we mean a function $u(x)$ which is harmonic in this neighbourhood except for the infinite point, and, when $|x| \to \infty$, which is bounded in the case $n = 2$ and tends to zero not more slowly than $|x|^{2-n}$ for $n > 2$ does.

Let D be a domain in E_n with a sufficiently smooth $(n-1)$-dimensional boundary $S = \partial D, u(x)$ and $v(x)$ be two given real functions, harmonic in D and having continuous derivatives of the first order on $D \cup S$ and derivatives of the second order which are integrable over D (the concept of a smooth boundary of a domain will be explained precisely in §3.1 of Chapter 3).

Integrating over the domain D the identities

$$\sum_{i=1}^{n} \frac{\partial}{\partial x_i}\left(u\frac{\partial u}{\partial x_i}\right) = \sum_{i=1}^{n}\left(\frac{\partial u}{\partial x_i}\right)^2$$

and

$$\sum_{i=1}^{n} \frac{\partial}{\partial x_i}\left(v\frac{\partial u}{\partial x_i} - u\frac{\partial v}{\partial x_i}\right) = 0$$

and making use of the Gauss-Ostrogradskii formula, we obtain respectively

$$\int_S u(y)\frac{\partial u(y)}{\partial \nu_y} ds_y = \sum_{i=1}^{n} \int_D \left(\frac{\partial u}{\partial x_i}\right)^2 d\tau_x \tag{83}$$

and

$$\int_S \left[v(y)\frac{\partial u}{\partial \nu_y} - u(y)\frac{\partial v}{\partial \nu_y} \right] ds_y = 0 \ . \tag{84}$$

When the domain D contains an infinite point of the space E_n, it is natural to require the intergrand expressions in the formulae (83) and (84) to be absolutely integrable (or integrable in Lebesgue's sense) for the formulae (83) and (84) to remain valid.

The formulae (83) and (84) permit us to establish easily the following important properties: (1) if a function $u(x)$ which is harmonic in D and satisfies certain conditions that are sufficient for the validity of the identity (83) is equal to zero at every point of S, then $u(x) = 0$ on all $D \cup S$; (2) of its normal derivative $\frac{\partial u}{\partial \nu}$ equals to zero at every point of S then $u(x) =$ constant on all $D \cup S$; (3) the integral over S of the normal derivative $\frac{\partial u}{\partial \nu}$ of a function $u(x)$ which is harmonic in D and satisfies certain conditions that are sufficient for the validity of the identity (84) equals to zero.

Since the function $u(x)$ is real, the correctness of statements (1) and (2) follows immediately from the identity (83), and the statement (3) will be verified if we set $v(x) \equiv 1$ for the whole domain D in the identity (84).

2.7.2. Let $r = |x - y|$, where x and y are arbitrary points in E_n. Then the function $u(r)$ depending only on r for $r \neq 0$ is a solution of Eq. (81) if and only if

$$\frac{d^2u}{dr^2} + \frac{n-1}{r}\frac{du}{dr} = 0 \ . \tag{85}$$

The ordinary differential equation (85) has two linearly independent solutions which can be written in the form

$$u_1(r) = 1 \ , \quad u_2(r) = \begin{cases} -\log\ r \ , & n = 2 \\ \dfrac{1}{r^{n-2}} \ , & n > 2 \ . \end{cases}$$

In the theory of harmonic functions the function

$$E(x,y) = \begin{cases} -\log\ |x-y| \ , & n = 2 \\ \dfrac{1}{(n-2)|x-y|^{n-2}} \ , & n > 2 \end{cases} \tag{86}$$

which is a solution of equation (81) in variables $x_1, \dots, x_n$ as well as in variables $y_1, \dots, y_n$ when $x \neq y$, plays an important role.

The function $E(x, y)$ defined by the formula (86) is called an *elementary* or a *fundamental solution* of Laplace's equation (81).

For a function $u(x)$ which is harmonic in a domain D with sufficiently smooth $(n-1)$-dimensional boundary S and has continuous derivatives of the first order on $D \cup S$ and integrable derivatives of the second order in D, the integral representation

$$u(x) = \frac{1}{\omega_n} \int_S E(x,y) \frac{\partial u(y)}{\partial \nu_y} ds_y - \frac{1}{\omega_n} \int_S u(y) \frac{\partial E(x,y)}{\partial \nu_y} ds_y \tag{87}$$

holds, where $E(x, y)$ is the solution (86) of Eq. (81), $\omega_n = \frac{2\pi^{\frac{n}{2}}}{\Gamma(\frac{n}{2})}$ is the area of a unit sphere in E_n, and Γ is *Euler's gamma function.*

To derive the formula (87) we exclude a point x in domain D along with a closed ball $|y - x| \leq \varepsilon$ that has a radius ε and belongs to D, and apply to the remaining part D_ε of the domain D, which is bounded by the surface S and the sphere $|y - x| = \varepsilon$, the formula (84) in which $v(y) = E(x, y)$. We then have

$$\begin{aligned}
&\int_S \left[E(x,y) \frac{\partial u(y)}{\partial \nu_y} - u(y) \frac{\partial E(x,y)}{\partial \nu_y} \right] ds_y \\
&\int_{|y-x|=\varepsilon} \left[E(x,y) \frac{\partial u(y)}{\partial \nu_y} - u(y) \frac{\partial E(x,y)}{\partial \nu_y} \right] ds_y \\
&= \int_{|y-x|=\varepsilon} E(x,y) \frac{\partial u(y)}{\partial \nu_y} ds_y - \int_{|y-x|=\varepsilon} [u(y) - u(x)] \frac{\partial E(x,y)}{\partial \nu_y} ds_y \\
&- u(x) \int_{|y-x|=\varepsilon} \frac{\partial E(x,y)}{\partial \nu_y} ds_y \ .
\end{aligned} \tag{88}$$

Taking into account the fact that on the sphere $|y - x| = \varepsilon$

$$E(x,y) = \begin{cases} -\log \, \varepsilon \ , & n = 2 \\ \dfrac{1}{(n-2)\varepsilon^{n-2}} \ , & n > 2 \ , \end{cases}$$

$$\frac{\partial E(x,y)}{\partial \nu_y} = -\frac{1}{\varepsilon^{n-1}} \ , \quad n \geq 2 \ ,$$

and

$$\lim_{\varepsilon\to 0}\int_{|y-x|=\varepsilon}[u(y)-u(x)]\frac{\partial E(x,y)}{\partial\nu_y}ds_y=0\ ,$$

$$\int_{|y-x|=\varepsilon}\frac{1}{\varepsilon^{n-1}}ds_y=\omega_n\ ,$$

we obtain the integral representation (87) by letting $\varepsilon\to 0$ and taking limit in (88) due to the property (3) of harmonic functions.

2.7.3. Suppose that the ball $|y-x|\le R$ is wholly in the domain of harmonicity of a function $u(x)$. Because of

$$E(x,y)=\begin{cases}-\log\ R\ , & n=2\\ \dfrac{1}{(n-2)R^{n-2}}\ , & n>2\end{cases}$$

and

$$\frac{\partial E(x,y)}{\partial\nu_y}=-\frac{1}{R^{n-1}}\ ,\quad n\ge 2$$

on the sphere $|y-x|=R$, from the formula (87) written for the ball $|y-x|<R$ and the property (3) of harmonic functions it follows that

$$u(x)=\frac{1}{\omega_n R^{n-1}}\int_{|y-x|=R}u(y)ds_y\ . \tag{89}$$

Rewriting the formula (89) for the sphere $|y-x|=\rho\le R$ in the form

$$\rho^{n-1}u(x)=\frac{1}{\omega_n}\int_{|y-x|=\rho}u(y)ds_y$$

and integrating with respect to ρ over the interval $0\le\rho\le R$, we obtain

$$u(x)=\frac{n}{\omega_n R^2}\int_{|y-x|\le R}u(y)d\tau_y\ , \tag{90}$$

where $d\tau_y$ is the volume element in the variable y and $\frac{\omega_n}{n}R^n$ is the volume of the ball $|y-x|<R$.

The formulae (89) and (90) are respectively known as the *formulae of arithmatical mean over a sphere* and that *over a ball* for a harmonic functions.

By M and m we denote the upper bound and the lower bound on a domain D of a harmonic function $u(x)$.

An *extremum principle of harmonic functions* follows from the formula (90): a non-constant valued harmonic function $u(x)$ defined in a domain D can take either the value M nor the value m at some point $x \in D$.

If $M = +\infty$ and $m = -\infty$, the correctness of this principle is clear due to the function $u(x)$ taking only a finite value at every point in domain D. If $M \neq \infty$, let us assume the inverse, i.e. $u(x_0) = M, x_0 \in D$, and consider in D a ball $|x - x_0| < \varepsilon$ for which $u(x) = M$ at every point. In fact, if $u(y) < M$ at some point y where $|y - x_0| < \varepsilon$ (the inequality $u(y) > M$ is excluded), then this inequality would be wholly kept on some neighbourhood of y due to the continuity of $u(x)$, and hence, we would have found a senseless inequality $M < M$ by using the formula (90) applied to the ball $|x - x_0| < \varepsilon$. Therefore, $u(x) = M$ on whole ball $|x - x_0| < \varepsilon$.

Now, let x be an arbitrary fixed point in the domain D and l be a continuous curve in D connecting x with x_0. The number ε can be taken as a smaller one than the distance between the boundary S of the domain D and the curve l. Shifting along the curve l the centre of the ball $|\eta - y| < \varepsilon$ from x_0 to x and making use of the fact that, for any of such $y, u = M$ in the interior of this ball, we obtain $u(x) = M$. Therefore, $u(x)$ is constant in D. The result at contradiction thus proves the correctness of the first part of stated assertion. The second part of it can be obtained from the first part if we consider $-u(x)$ instead of $u(x)$.

If it is known, in addition, that the non-constant function $u(x)$ which is harmonic in the domain D is continuous on $D \cup S$, then from the proved extremum principle it follows immediately that the maximum point and minimum point of this function are all on the boundary S of the domain D.

2.7.4. Let $f(x)$ be a real function given on the $(n-1)$-dimensional boundary S of the domain D. The problem of determining the function which is harmonic in the domain D and satisfies the boundary condition

$$\lim_{\substack{x \to x_0 \\ x \in D}} u(x) = f(x_0) , \quad x_0 \in S , \tag{91}$$

is called a *first boundary-value problem* or a *Dirichlet problem of Laplace's equation* (81).

Under the further requirement that the function $f(x)$ be continuous on S and the desired solution $u(x)$ be continuous on $D \cup S$, the problem (81)–(91) has no more than one solution (the uniqueness theorem of solution for Dirichlet's problem).

By virtue of the linearity of equation (81) and the condition (91), the validity of this assertion will be verified if it can be proved that when $f(x)$ is identically equal to zero the problem (81)–(91) has no solution but the one that equals to zero identically on $D \cup S$. In turn, the last statement is an immediate consequence of the extremum principle of harmonic functions that has just been proved above.

Requiring the continuity of the function $f(x)$, it can also be shown, by using the same extremum principle, that a small variation of $f(x)$ will bring out only a small variation in a solution on $D \cup S$ for the Dirichlet problem.

2.7.5. We suppose that the boundary S of the domain D is a sufficiently smooth $(n-1)$-dimensional surface, and the desired solution of problem (81)–(91) has first order derivatives which are continuous on $D \cup S$ and second order derivatives which are integrable over D. Under these assumptions the integral representation (87) for the function $u(x)$ holds.

The function $G(x,y)$ of two points x and y which is defined on $D \cup S$ and has the properties: (1) it has the form

$$G(x,y) = E(x,y) + g(x,y) \ , \tag{92}$$

where $E(x,y)$ is the elementary solution of Eq. (81) and $g(x,y)$ is a harmonic function in x as well as in y for $x, y \in D$; (2) when at least one of points x, y is on S, then

$$G(x,y) = 0 \ ; \tag{93}$$

is called a *Green's function* of the problem (81)–(91).

For finding a Green's function $G(x,y)$, taking (92) and (93) into account it should be appropriate to construct a solution of the Dirichlet problem for the function $g(x,y)$ which is harmonic in D, when function $f(x_0)$ given on S in the boundary condition (91) has a specific form, i.e., $f = g(x_0, y) = -E(x_0, y)$.

Under the condition that S is sufficiently smooth, such a solution $g(x, y)$ for this problem must exist and it behaves in such a way that the formulae (84) and (87) remain valid.

It follows immediately from the definition of Green's function and the extremum principle of harmonic functions that $G(x, y) \geq 0$ on whole $D \cup S$.

The Green's function $G(x, y)$ is symmetric in x, y. To show this we remove from D two points x and y along with two balls $d : |z-x| \leq \delta$ and $d' : |z-y| \leq \delta$ of sufficiently small radius δ and denote by D_δ the remaining part of the domain D.

Applying the formula (84) to the domain D_δ and writing $u(z) = G(z, x)$, $v(z) = G(z, y)$, we find that

$$\int_S \left[G(z,y)\frac{\partial G(z,x)}{\partial \nu_z} - G(z,x)\frac{\partial G(z,y)}{\partial \nu_z}\right] ds_z$$
$$= \left(\int_C + \int_{C'}\right) \left[G(z,y)\frac{\partial G(z,x)}{\partial \nu_z} - G(z,x)\frac{\partial G(z,y)}{\partial \nu_z}\right] ds_z \ ,$$

where ν_z is the outward normal at a point z of S or sphere $C : |z - x| = \delta$ or $C' : |z - y| = \delta$. Since $G(z, x) = G(z, y) = 0$ for $z \in S$, we can rewrite this formula in the form

$$\int_C \left[G(z,y)\frac{\partial G(z,x)}{\partial \nu_z} - G(z,x)\frac{\partial G(z,y)}{\partial \nu_z}\right] ds_z$$
$$= \int_{C'} \left[G(z,x)\frac{\partial G(z,y)}{\partial \nu_z} - G(z,y)\frac{\partial G(z,x)}{\partial \nu_z}\right] ds_z$$

where ν_z is the outward normal at a point z of S or sphere $C' : |z - x| = \delta$ or $C' : |z - y| = \delta$. Since $G(z, x) = G(z, y) = 0$ for $z \in S$, we can rewrite this formula in the form

$$\int_C \left[G(z,y)\frac{\partial G(z,x)}{\partial \nu_z} - G(z,x)\frac{\partial G(z,y)}{\partial \nu_z}\right] ds_z$$
$$= \int_{C'} \left[G(z,x)\frac{\partial G(z,y)}{\partial \nu_z} - G(z,y)\frac{\partial G(z,x)}{\partial \nu_z}\right] ds_z \ .$$

From this and taking into account

$$G(z,x) = E(z,x) + g(z,x) \ , \quad G(z,y) = E(x,y) + g(z,y) \ ,$$

where $g(z,x)$ and $g(z,y)$ are harmonic functions, by letting $\delta \to 0$ just as in the deduction of the formula (87) we come to the conclusion that $G(x,y) = G(y,x)$, which was to be proved.

The integral representation (87) obviously remains valid if we use a Green's function $G(x,y)$ instead of $E(x,y)$:

$$u(x) = -\frac{1}{\omega_n} \int_S \frac{\partial G(x,y)}{\partial \nu_y} f(y) ds_y \ . \tag{94}$$

The harmonicity of the function $u(x)(x \in D)$ represented by the formula (94) follows from the harmonicity of the Green's function $G(x,y)$ in x for $x \neq y$. Thus, what we need to prove specially is that this function also satisfies boundary condition (91). Therefore the formula (94) gives a solution for problem (81)–(91) in the presence of a Green's function for the domain D.

2.7.6. We now show that for $|x| < 1$ a Green's function $G(x,y)$ is given by the formula

$$G(x,y) = E(x,y) - E\left(|x|y \, , \, \frac{x}{|x|}\right) . \tag{95}$$

In fact, since

$$\begin{aligned} \left| |x|y - \frac{x}{|x|} \right| &= \left[|x|^2|y|^2 - 2xy + 1 \right]^{\frac{1}{2}} = \left| |y|x = \frac{y}{|y|} \right| \\ &= |y| \left| x - \frac{y}{|y|^2} \right| = |x| \left| y - \frac{x}{|x|^2} \right| , \end{aligned}$$

the function

$$g(x,y) = -E\left(|x|y \, , \, \frac{x}{|x|}\right)$$

for $|x| < 1$ and $|y| < 1$ is harmonic in x as well as in y. Moreover, in the presence of at least one of the equalities $|x| = 1$ and $|y| = 1$, say $|y| = 1$, we have

$$\begin{aligned} |y - x| &= \left[|x|^2 - 2xy + 1 \right]^{\frac{1}{2}} = \left| |y|x - \frac{y}{|y|} \right| \\ &= \left| |x|y - \frac{x}{|x|} \right| . \end{aligned}$$

Thus, the function $G(x, y)$ defined by the formula (95) satisfies all the requirements pertaining to a Green's function.

Since for $|y| = 1$,

$$\frac{\partial G(x,y)}{\partial \nu_y} = -\sum_{i=1}^{n}\left\{\frac{y_i(y_i - x_i)}{|y-x|^n} - |x|\frac{y_i\left(|x|y_i - \frac{x}{|x|}\right)}{\left||x|y - \frac{x}{|x|}\right|^n}\right\}$$

$$= -\frac{1-|x|^2}{|y-x|^n},$$

The formula (94) in the considered case can be written as

$$u(x) = \frac{1}{\omega_n}\int_{|y|=1}\frac{1-|x|^2}{|y-x|^n}f(y)ds_y . \tag{96}$$

This formula is called *Poisson's formula.*

By direct check it is not difficult to become convinced that the harmonic function $u(x)$ defined on the ball $|x| < 1$ by Poisson's formula (96) satisfies the boundary condition (91) on the requirement that the function $f(x)$ is continuous.

For the sake of clearness of the reasoning we confine ourselves to considering the case $n = 2$, that is,

$$u(x) = \frac{1}{2\pi}\int_0^{2\pi}\frac{1-|x|^2}{|y-x|^2}f(y)d\psi ,$$
$$y_1 = \cos\psi , \quad y_2 = \sin\psi , \quad y = (y_1, y_2) . \tag{97}$$

Since $u_0(x) \equiv 1$ is a harmonic function satisfying the boundary condition $\lim\limits_{x\to x_0} u_0(x) = 1$, $|x| < 1$, where $x_0 = (x_{10}, x_{20})$ is an arbitrary fixed point on the circumference $|x_0| = 1$, therefore, for all points in the disk $|x| < 1$ we have

$$u_0(x) = \frac{1}{2\pi}\int_0^{2\pi}\frac{1-|x|^2}{|y-x|^2}d\psi \tag{98}$$

from the formula (97).

On the basis of (97) and (98) we obtain

$$u(x) - f(x_0) = \frac{1}{2\pi}\int_0^{2\pi}\frac{1-|x|^2}{|y-x|^2}\left[f(y) - f(x_0)\right]d\psi . \tag{99}$$

Since the function $f(x)$ is uniformly continuous on the circumference $|x| = 1$, for any preassigned $\varepsilon > 0$ there is a number $\delta(\varepsilon) > 0$ such that, for all ψ, ψ_0, $x_{10} = \cos\ \psi_0$ and $x_{20} = \sin\psi_0$ which satisfies the condition $|\psi - \psi_0| < \delta$, the inequality

$$|f(y) - f(x_0)| < \varepsilon \tag{100}$$

holds.

We rewrite the expression (99) in the form $u(x) - f(x_0) = I_1 + I_2$, where

$$I_1 = \frac{1}{2\pi}\int_{\psi_0-\delta}^{\psi_0+\delta} \frac{1-|x|^2}{|y-x|^2}[f(y) - f(x_0)]d\psi \ ,$$

$$I_2 = \frac{1}{2\pi}\left(\int_0^{\psi_0-\delta} + \int_{\psi_0+\delta}^{2\pi}\right) \frac{1-|x|^2}{|y-x|^2}[f(y) - f(x_0)]d\psi \ .$$

On the basis of (98) and (100) we arrive at the conclusion that $|I_1| < \varepsilon$.

After choosing a $\delta(\varepsilon), x$ can be taken to be closed to x_0 sufficiently so that the inequality

$$\left(\int_0^{\psi_0-\delta} + \int_{\psi_0+\delta}^{2\pi}\right) \frac{1-|x|^2}{|y-x|^2}d\psi < \frac{\pi\varepsilon}{M} \ , \qquad M = \max|f(y)|, \ |y| = 1$$

i.e., $|I_2| < \varepsilon$, holds. Therefore $|u(x) - f(x_0)| < 2\varepsilon$ and hence

$$\lim_{x\to x_0} u(x) = f(x_0) \ , \quad |x| < 1 \ , \quad |x_0| = 1 \ .$$

Thus, Poisson's formula (97) gives a solution of the Dirichlet problem formulated as follows: to find a function $u(x)$ which is harmonic in disk $|x| < 1$ and continuous on the closed disk $|x| \leq 1$ and satisfies the boundary condition

$$u(x_0) = f(x_0)$$

on the whole circumference $|x_0| = 1$.

In what follows we shall verify that under some sufficiently general assumptions concerning the domain D and the function $f(x)$ given on its boundary S, the problem (81)–(91) has a stable solution, and it is a unique one; that is, this problem is well-posed.

On the basis of the formula (96) and the properties (b) and (c) of harmonic functions, it is easy to see that if the function $f(x)$ is continuous then a solution of the problem (81)–(91) in the ball $|x - x_0| < R$ is given by the formula

$$u(x) = \frac{1}{\omega_n R} \int_{|y-x_0|=R} \frac{R^2 - |x - x_0|^2}{|y - x|^n} f(y) ds_y \ , \tag{101}$$

and the formula

$$u(x) = \frac{1}{\omega_n R} \int_{|y-x_1|=R} \frac{|x - x_0|^2 - R^2}{|y - x|^n} f(y) ds_y \tag{102}$$

gives a solution for the Dirichlet problem of harmonic functions, which is formulated as follows: to find a function $u(x)$ which is regular and harmonic outside the closed ball $|x - x_0| \leq R$, continuous outside the ball $|x - x_0| < R$, satisfying the boundary condition (91) on the sphere $|y - x_0| = R$ (the exterior Dirichlet problem of harmonic functions).

2.7.7. In the case that D is the half-space $x_n > 0$, a Green's function $G(x, y)$ is also constructed in an explicit form. In fact, let x and y be two points of this half-space, $y' = (y_1, \ldots, y_{n-1}, -y_n)$ with respect to the plane $y_n = 0$. Since for $x_n > 0$ and $y_n > 0$, the function $g(x, y) = -E(x, y')$ is harmonic in x as well as in y and moreover $E(x, y) - E(x, y') = 0$ for $y_n = 0$,

$$G(x, y) = E(x, y) - E(x, y') \tag{103}$$

is a Green's function for the considered half-space.

Suppose that the desired solution $u(x)$ for the Dirichlet problem (81)–(91), in the considered case, vanishes as $|x| \to \infty$ in such a way that the formula (94) is also valid this time. For instance, this is trivial if as $|x| \to \infty$

$$|u(x)| < \frac{A}{|x|^h} \ , \quad \left| \frac{\partial u}{\partial x_i} \right| < \frac{A}{|x|^{h+1}} \ , \quad i = 1, \ldots, n \ ,$$

where A and h are positive constants. According to this, for large $\delta = (\sum_{i=1}^{n-1} y_i^2)^{\frac{1}{2}}$ the function $f(y_1, \ldots, y_{n-1})$ given on the plane $y_n = 0$ should satisfy the condition

$$|f| < \frac{A}{\delta^h} \ .$$

Substituting the expression (103) of $G(x, y)$ into the right-hand side of the formula (94) and taking into account

$$\frac{\partial G(x,y)}{\partial \nu_y} = -\frac{\partial G(x,y)}{\partial y_n} = \frac{y_n - x_n}{|y-x|^n} - \frac{y_n + x_n}{|y'-x|^n}$$
$$= -\frac{2x_n}{\left[\sum_{i=1}^{n-1}(y_i - x_i)^2 + x^2\right]^{\frac{h}{2}}}$$

when $y_n = 0$, we obtain the formula

$$u(x) = \frac{\Gamma\left(\frac{n}{2}\right)x_n}{\pi^{\frac{n}{2}}} \int_{y_n=0} \frac{f(y_1, \ldots, y_{n-1})}{\left[\sum_{i=1}^{n-1}(y_i - x_i)^2 + x_n^2\right]^{\frac{n}{2}}} dy_1 \ldots dy_{n-1} , \tag{104}$$

which gives a solution for the Dirichlet problem of functions which are harmonic on the half-plane $x_n > 0$ and satisfy the boundary condition

$$\lim_{x \to y} u(x) = f(y_1, \ldots, y_{n-1}) , \quad x_n > 0 , \quad y_n = 0 . \tag{105}$$

The formulae (101), (102) and (104) are also called *Poisson's formulae.*

The formula (104) can also be used for the case when the function $f(y)$ is continuous, and bounded if $y_n = 0$. We can also start from this to check the boundary condition (105) as we had done for the Dirichlet problem on a disk.

2.7.8. From the formula (101) applied to the ball $|x| < R$,

$$u(x) = \frac{1}{\omega_n R} \int_{|y|=R} \frac{R^2 - |x|^2}{|y-x|^n} u(y) ds_y , \tag{106}$$

we can show the correctness of the following assertion: if a function $u(x)$ which is harmonic in the space E_n is nonnegative (nonpositive) everywhere, then it must be a constant.

In fact, because for $|x| < R, |y| = R$ the inequalities $R - |x| < |y - x| < R + |x|$ hold and $u(x) \geq 0$ by assumption, from (106) and (89) we have

$$R^{n-2} \frac{R - |x|}{(R + |x|)^{n-1}} u(0) \leq u(x) \leq R^{n-2} \frac{R + |x|}{(R - |x|^{n-1}} u(0) \tag{107}$$

for any $R > 0$. Thus, arbitrarily fixing a point $x \in E_n$ and then letting R tends to the infinity, we find that the function $u(x) = u(0)$ at every point x in the space E_n.

In turn, from this conclusion the following *Liouville's theorem* can be proved: if a function $u(x)$ which is harmonic in E_n is upper (lower) bounded, then it is a constant.

Indeed, suppose that $u(x) < M$ for all $x \in E_n$, where M is a constant. Because the function $M - u(x)$ is harmonic in E_n and nonnegative, $M - u(x) = M - u(0)$, i.e., $u(x) = u(0)$ as it has just been proved.

2.7.9. Liouville's theorem makes it possible to conclude that the Dirichlet problem considered above for the half space $x_n > 0$ in the class of bounded harmonic functions which are continuous up to $x_n = 0$ has no more than one solution.

In fact, the difference $u(x) = u_1(x) - u_2(x)$ between any two bounded solutions $u_1(x)$ and $u_2(x)$ of this problem satisfies the boundary condition $u(x) = 0$ when $x_n = 0$. Consider the function

$$v(x) = \begin{cases} u(x_1, \ldots, x_{n-1}, x_n), & x_n \geq 0 \\ -u(x_1, \ldots, x_{n-1}, -x_n), & x_n \leq 0, \end{cases}$$

which is harmonic for $x_n > 0$ as well as for $x_n < 0$. The function $v(x)$ is harmonic on the whole space E_n because in the ball $|x| < R$ for any $R > 0$ it coincides with a harmonic function $w(x)$ that satisfies the boundary condition $w(x) = v(x)$ for $|x| = R$. Since $v(x)$ is bounded by virtue of Liouville's theorem $w(x) =$ constant in E_n. But $w(x) = 0$ for $x_n = v$, so we have $w(x) = 0$ in all E_n and, therefore, $u_1(x) = u_2(x)$.

2.7.10. It is well known that, with respect to a given function $u(x_1, x_2)$ which is harmonic in a simply connected domain D of the complex variable $z = x_1 + ix_2$, we can, up to an arbitrary imaginary constant term, change it into a function $F(z)$ which is analytic in D and whose real part is $u(x_1, x_2)$.

In the final part of this section we will show that, for the case of the disk $D = \{|z| < 1\}$ and under the assumption that $u(x_1, x_2)$ is continuous on $D \cup S$, $S = \{|z| = 1\}$, Poisson's formula (97) makes it possible to find the function $F(z)$ in terms of the values of $u(y_1, y_2)$ on the circumference S.

In fact, by using the notations $u(x) = u(z), z = x_1 + ix_2$, $f(y) = \tilde{f}(\psi) =$

$u(t), t = e^{i\psi} = y_1 + iy_2$, $dt = itd\psi$, we can rewrite the formula (97) in the form

$$u(z) = \frac{1}{2\pi}\int_S \frac{1-|z|^2}{|t-z|^2}\frac{u(t)dt}{it} \tag{106'}$$

In view of the equality

$$1-|z|^2 = \mathrm{Re}(t+z)(\bar{t}-\bar{z}) , \quad |t-z|^2 = (t-z)(\bar{t}-\bar{z}) ,$$

we obtain from (106)′

$$\begin{aligned} u(z) &= \mathrm{Re}\frac{1}{2\pi i}\int_S \frac{t+z}{t(t-z)}u(t)dt \\ &= \mathrm{Re}\frac{1}{\pi i}\int_S \left(\frac{1}{t-z} - \frac{1}{2t}\right) u(t)dt . \end{aligned}$$

Thus, for the function $F(z)$ we can take the expression

$$F(z) = \frac{1}{\pi i}\int_S \left(\frac{1}{t-z} - \frac{1}{2t}\right) u(t)dt + iC , \tag{107'}$$

where C is an arbitrary real constant.

The formula (107)′ is usually called *Schwarz's formula.*

§2.8 Heat-conduction Equation

2.8.1. Since corresponding to the heat-conduction equation

$$\frac{\partial u}{\partial t} - \sum_{i=1}^{n}\frac{\partial^2 u}{\partial x_i^2} = 0 \tag{108}$$

the characteristic form $Q(\lambda_0, \lambda_1, \dots, \lambda_n)$ has the canonical form

$$Q = -\sum_{i=1}^{n}\lambda_i^2$$

with deficiency index 1, it is a parabolic type equation.

By direct computation it can be shown that the function defined by

$$u(x,t) = \sum_{j=0}^{\infty}\frac{t^j}{j!}\Delta^j\tau(x_1,\dots,x_n) , \quad \Delta = \sum_{i=1}^{n}\frac{\partial^2}{\partial x_i^2} , \tag{109}$$

where τ is any given infinitely differentiable function of variables $x_1, \dots, x_n$, satisfies Eq. (108), provided that the series on the right-hand side of (109) and the series obtained from it by termwise differentiation once in t and twice in $x_i, i = 1, \dots, n$, are uniformly convergent.

The function

$$E(x, y; t, t_0) = (t - t_0)^{-\frac{n}{2}} \exp\left[-\frac{1}{4(t-t_0)}\sum_{i=1}^{n}(x_i - y_i)^2\right] , \qquad (110)$$

where $x = (x_1, \ldots, x_n)$ and $y = (y_1, \ldots, y_n)$ are points in the space E_n and the independent variables t and t_0 play the role of the time such that $t > t_0$, is also a solution of Eq. (108).

In fact,

$$\frac{\partial^2 E}{\partial x_i^2} = -\frac{1}{2(t-t_0)}E + \frac{(x_i - y_i)^2}{4(t-t_0)^2}E , \quad i = 1, \ldots, n ,$$

$$\frac{\partial E}{\partial t} = -\frac{n}{2(t-t_0)}E + \frac{E}{4(t-t_0)^2}\sum_{i=1}^{n}(x_i - y_i)^2 ,$$

and it follows from these that

$$\frac{\partial E}{\partial t} - \sum_{i=1}^{n}\frac{\partial^2 E}{\partial x_i^2} = 0 .$$

The function $E(x, y; t, t_0)$ defined by the formula (110) is called an elementary or fundamental solution of Eq. (108).

2.8.2. By Q we denote a domain in the space E_{n+1} of points $(x_1, \ldots, x_n, t)$ such that a domain δ in the plane $t = 0$ is used for its lower basis, and that its lateral surface is cylinder S_0 with elements parallel to the axis t and $t \geq 0$.

Let h be an arbitrary positive number. We denote the part with $0 < t < h$ of the domain Q by Q_h, its lateral surface by S_h, the union $S_h \cup \delta$ by S and the upper basis of Q_h, which is a domain in the plane $t = h$, by δ_h.

For the solution of Eq. (108) the following *extremum principle* is valid: a solution $u(x, t)$ of equation (108), which is regular in the domain Q and continuous in $Q \cup S_0 \cup \delta$, reaches (on S) its extremum in $Q_h \cup \partial Q_h$.

In fact, we denote by M the maximum of $u(x, t)$ on the closed set $Q_h \cup \partial Q_h$. Let us assume that the function $u(x, t)$ reaches this maximum at some point $(x_0, t_0) \in Q_h \cup \delta_h$ and that $M = u(x_0, t_0) > u(x, t)$ for all $(x, t) \in S$. This assumption will lead to some contradiction.

Consider a function

$$v(x, t) = u(x, t) + a(h - t) , \qquad (111)$$

where a is a positive constant. Since $o \leq t \leq h$, we have from (111)

$$u(x,t) \leq v(x,t) \leq u(x,t) + ah \tag{112}$$

on all $Q_h \cup \partial Q_h$.

We denote by M_u^s and M_v^s the maxima of $u(x,t)$ and $v(x,t)$ on S respectively. $M_u^s < M$ by the assumption. We choose the number a such that the equality

$$a < \frac{M - M_u^s}{h} \tag{113}$$

holds. On the basis of (112) and (113) we obtain

$$\begin{aligned} M_v^s \leq M_u^s + ah &< M_u^s + \frac{M - M_u^s}{h} h \\ &= M = u(x_0, t_0) \ . \end{aligned}$$

From here it follows that the function $v(x,t)$ cannot reach its maximum on S. Therefore this function reaches its maximum in $Q_h \cup \partial Q_h$ at some point $(x', t') \in Q_h \cup \delta_h$.

At first we assume that $(x', t') \in Q_h$. Because (x', t') is a maximum point of the function $v(x,t)$ on $Q \cup \partial Q_h$,

$$\frac{\partial v}{\partial t} = 0 \ , \quad \sum_{i=1}^{n} \frac{\partial^2 v}{\partial x_i^2} \leq 0$$

at this point, i.e.,

$$\frac{\partial v}{\partial t} - \sum_{i=1}^{n} \frac{\partial^2 v}{\partial x_i^2} \geq 0 \ . \tag{114}$$

Now suppose $(x', t') \in \delta_h$, i.e., $(x', t') = (x', h)$. Since $v(x,t)$ reaches its maximum in $Q_h \cup \partial Q_h$ at the point (x', h), we have

$$\frac{\partial v}{\partial t} \geq 0 \tag{115}$$

at this point. Taking into account the fact that (x', h) is a maximum point of $v(x,h)$ as a function of x in the domain δ_h, we should have

$$\sum_{i=1}^{n} \frac{\partial^2 v(x', h)}{\partial_i x'^2} \leq 0 \ . \tag{116}$$

By virtue of (115) and (116), we come again to the estimate (114) at the point $(x',t') = (x',h)$. Substituting for $\frac{\partial v}{\partial t}$ and $\frac{\partial^2 v}{\partial x_i^2}, i = 1,\dots,n$, the values obtained from (111) on the left-hand side of (114), we find

$$\frac{\partial u}{\partial t} - a - \sum_{i=1}^{n} \frac{\partial^2 u}{\partial x_i^2} \geq 0 ,$$

i.e.,

$$\frac{\partial u}{\partial t} - \sum_{i=1}^{n} \frac{\partial^2 u}{\partial x_i^2} > 0$$

at the point (x',t'). But this contradicts the assumption that $u(x,t)$ is a solution of Eq. (108).

The proof for past two of the stated assertion can be obtained by using $-u(x,t)$ instead of $u(x,t)$ in the above.

In the statement of the extremum principle the requirement that the lateral surface S_0 of the domain Q be cylinderical is not necessary at all. From the proof mentioned above it follows that this principle remains valid under certain more general assumptions concerning S_0.

2.8.3. As shown in §1.4 of Chapter 1, under certain assumptions the phenomenon of heat propagation in a medium is characterized by equation (108) with $n = 1,2,3$.

Denote by Q_T the domain of variables (x,t),

$$Q_T = D \times \{t_0 < t < T\} .$$

In observing the heat propagation it is possible, evidently, to evaluate the values of the function $u(x,t)$ at every point x of the heat-conducting medium at the starting instant t_0 and at every point x_0 of the boundary ∂D for all the values t of time in the interval $t_0 < t < T$.

Therefore we can assume that the values of the desired solution $u(x,t)$ of equation (108) in Q_T are preassigned on the lower basis D of the domain Q_T and on its lateral surface $\partial D \times \{t_0 < t < T\}$:

$$u(x,t_0) = \varphi(x) , \quad x \in D , \tag{117}$$

$$u(x_0,t) = \psi(t) \ , \quad (x_0,t) \in \partial D \times \{t_0 < t < T\} \ . \tag{118}$$

The conditions (117) and (118) are usually called an ***initial condition*** and a ***boundary condition***, respectively. The problem to search for a solution $u(x,t)$ of Eq. (108) which is regular in Q_T and satisfies the conditions (117) and (118) is called a ***first boundary value problem***.

By virtue of the linearity of Eq. (108) and the conditions (117) and (118), on the basis of the extremum principle we can conclude at once that the problem (108), (117), (118) cannot have more than one solution which is continuous on $Q_T \cup \partial Q_T$. The stability of the solution for this problem also follows from the extremum principle.

2.8.4. The existence of solutions for problem (108), (117), (118) has been proved under quite general conditions concerning the boundary ∂D of the domain D and the functions $\varphi(x)$ and $\psi(t)$. Here we confine ourselves to considering the case that (1) the number of space variables in equation (108) is equal to 1, that is,

$$\frac{\partial u}{\partial t} - \frac{\partial^2 u}{\partial x^2} = 0 \ ; \tag{119}$$

(2) the domain Q_T is the rectangle

$$O(0,0) \ , \quad B(0,T) \ , \quad N(l,T) \ , \quad A(l,0) \ , \quad l > 0 \ , \quad t > 0 \ ;$$

(3) the conditions (118) and (117) have the form

$$u(0,t) = u(l,t) = 0 \ , \quad 0 \le t \le T \ , \tag{120}$$

$$u(x,0) = \varphi(x) \ , \quad 0 \le x \le l \ , \tag{121}$$

and furthermore, $\varphi(x)$ is continuously differentiable and vanishes at $x = 0$ and $x = l$.

As it is well known, the function $\varphi(x)$ can be expanded into an absolutely and uniformly convergent Fourier series

$$\varphi(x) = \sum_{k=1}^{\infty} a_k \sin \frac{\pi k}{l} x \ , \tag{122}$$

where

$$a_k = \frac{2}{l}\int_0^l \varphi(x)\sin\frac{\pi k}{l}x dx\ , \quad k = 1, 2, \ldots .$$

Making use of the formula (109) with $n = 1, x_1 = x$ and $\tau(x) = \sin\frac{\pi k}{l}x$, we obtain the regular solution $u_k(x,t)$ of equation (119):

$$u_k(x,t) = e^{-\frac{\pi^2 k^2}{l^2}t}\sin\frac{\pi k}{l}x\ ,$$

which satisfies the boundary and initial conditions

$$u_k(0,t) = u(l,t) = 0\ , \quad u_k(x,0) = \sin\frac{\pi k}{l}x\ .$$

It is easy to show that a function $u(x,t)$ which can be expressed as the sum of the series

$$u(x,t) = \sum_{k=1}^{\infty} a_k e^{-\frac{\pi^2 k^2}{l^2}t}\sin\frac{\pi k}{l}x \tag{123}$$

is the desired solution of problem (119), (120), (121). For this it is enough to note that for $t > 0$ the series (123), and the series obtained from it by termwise differentiation in x and t for any number of times being absolutely and uniformly convergent in some neighbourhood of every point $(x,t) \in Q$ follows from the equality

$$\lim_{k\to\infty}\left(\frac{\pi k}{l}\right)^m e^{-\frac{\pi^2 k^2}{l^2}t} = 0\ , \quad m = 0, 1, \ldots .$$

Thus, the problem (119), (120), (122) is well-posed.

2.8.5. For the heat-conducting equation there are other problems which are also posed in addition to the first boundary value problem (117)–(118). Let us turn our attention to one of them now.

On the plane of variables x, t we take a zone $D : \{-\infty < x < \infty,\ 0 \le t < T\}$, where T is a fixed positive number, not excluding $T = \infty$.

By a regular solution in D of Eq. (119) we mean a function $u(x,t)$ which is bounded and continuous in D, having continuous derivatives u_t and u_{xx} and satisfying this equation in the whole interior of D.

The problem of determining in D a regular solution $u(x,t)$ of equation (119), which satisfies the condition

$$u(x,0) = \varphi(x) , \quad -\infty < x < \infty , \tag{124}$$

with $\varphi(x)$ being a given bounded continuous function, is said to be a *Cauchy-Dirichlet problem.*

Now we shall show that the function defined by the formula

$$u(x,t) = \frac{1}{2\sqrt{\pi t}} \int_{-\infty}^{\infty} \varphi(\xi) e^{-\frac{(\xi - x)^2}{4t}} d\xi \tag{125}$$

is a solution of the Cauchy-Dirichlet problem (119)–(124).

First of all note that in a neighbourhood of every interior point (x,t) of the zone D the integral on the right-hand side of (125) converges uniformly. Next, as a result of the change of variable $\xi = x + 2\eta\sqrt{t}$ the formula (125) takes the form

$$u(x,t) = \frac{1}{\sqrt{\pi}} \int_{-\infty}^{\infty} \left(x + 2\eta\sqrt{t}\right) e^{-\eta^2} d\eta . \tag{126}$$

Since $\sup |\varphi(x)| \leq M$, where M is a positive number, and the integral on the right-hand side of (126) converges absolutely,

$$|u(x,t)| \leq \frac{M}{\sqrt{\pi}} \int_{-\infty}^{\infty} e^{-\eta^2} d\eta .$$

From this and by taking into account

$$\int_{-\infty}^{\infty} e^{-\eta^2} = \sqrt{\pi} , \tag{127}$$

we obtain

$$|u(x,t)| \leq M . \tag{128}$$

In view of the continuity of the function $\varphi(x)$, the continuity of $u(x,t)$ in D follows from (126).

Taking into account the fact that the integrals, obtained by bringing the differential operations in x and t (any number of times) under the integral symbol on the right-hand side of (125), converge uniformly near every point $(x,t), t>0$, and the function

$$\frac{1}{\sqrt{t}}e^{-\frac{(\xi-x)^2}{4t}} = E(x,\xi,t,0)$$

is an elementary solution of equation (119), we conclude that the function $u(x,t)$ defined by the formula (126) is a solution of Eq. (119) at every interior point (x,t) of zone D.

Taking the limits as $t \to 0$ (this operation is legitimate because of the uniform convergence of the integral near every point $(x,0)$ when $t \geq 0$) we have from (126)

$$\lim_{t\to 0} u(x,t) = \varphi(x)$$

in view of (127). Thus, the proof of the existence of solutions for the Cauchy-Dirichlet problem (119)–(124) is complete.

2.8.6. The uniqueness and stability of the solution for this problem are an immediate consequence of the following conclusion (the *extremum principle for a zone*): For a regular solution in the zone D of Eq. (119), $u(x,t)$ the estimate

$$m \leq u(x,t) \leq M \tag{129}$$

holds, where

$$m = \inf\ u(x,0)\ , \quad M = \sup u(x,0)\ , \quad -\infty < x < \infty\ .$$

To determine the validity of the first estimate in (129), let us introduce for discussion a function $v(x,t) = x^2 + 2t$ which is a solution of Eq. (119).

By n we denote a lower bound of $u(x,t)$, where $(x,t) \in D$, and consider the function

$$w(x,t) = u(x,t) - m + \varepsilon\frac{v(x,t)}{v(x_0,t_0)}\ , \tag{130}$$

where ε is an arbitrary positive number and (x_0,t_0) is an arbitrary and fixed interior point of D.

Evidently, the function $w(x,t)$ represented by the formula (130) is a solution of Eq. (119) in the zone D. Moreover,

$$w(x,0) = u(x,0) - m + \varepsilon \frac{x^2}{x_0^2 + 2t_0} \geq 0 \tag{131}$$

for $t = 0$ and

$$w(x,t) \geq u(x,t) - n \geq 0 \tag{132}$$

for $|x| = |x_0| + \sqrt{\frac{(m-n)v(x_0,t_0)}{\varepsilon}}$.

From the estimates (131)–(132) and on the basis of the extremum principle proved above for solutions of the heat-conduction equation applied to the rectangular

$$0 \leq t \leq T \, ,$$

$$- |x_0| - \sqrt{\frac{(m-n)v(x_0,t_0)}{\varepsilon}} \leq x \leq |x_0| + \sqrt{\frac{(m-n)v(x_0,t_0)}{\varepsilon}}$$

which contains the point (x_0, t_0), we conclude that

$$w(x_0,t_0) = u(x_0,t_0) - m + \varepsilon \geq 0 \, ,$$

i.e., $u(x_0,t_0) \geq m - \varepsilon$. From this, in turn, due to the arbitrariness of ε it follows that, $u(x_0,t_0) \geq m$. Thus, $u(x,t) \geq m$ in the whole D.

Using $-u(x,t)$ in place of $u(x,t)$ in the above and repeating the argument, the second estimate in (129) can be verified.

2.8.7. The formula (125) which gives the solution of the Cauchy-Dirichlet problem makes it possible to write the solution as a integral of a solution of the homogeneous problem

$$u(x,0) = 0 \, , \quad -\infty < x < \infty \, , \tag{133}$$

for the *non-homogeneous heat-conduction equation*

$$u_t - u_{xx} = g(x,t) \, , \tag{134}$$

where $g(x,t)$ is a bounded continuous function given in the zone D.

In fact, for the carrier of given data in the Cauchy-Dirichlet problem considered above we take the straight line $t = \tau$, instead of $t = 0$, where τ is a fixed positive number, and in the zone $D_\tau : \{\tau \leq t \leq T, -\infty < x < \infty\}$ consider the problem

$$v_t - v_{xx} = 0 , \quad (x,t) \in D_\tau , \tag{135}$$

$$v = g(x,\tau) , \quad t = \tau . \tag{136}$$

The solution of problem (133)–(134) clearly is the function

$$v(x,t,\tau) = \frac{1}{2\sqrt{\pi}\sqrt{t-\tau}} \int_{-\infty}^{\infty} e^{-\frac{(\xi-x)^2}{4(t-\tau)}} g(\xi,\tau) d\xi .$$

On the basis of (135) and (136) we conclude that the function

$$u(x,t) = \int_0^t v(x,t,\tau) d\tau$$

is the solution of problem (134)–(133).

In the statement of the Cauchy-Dirichlet problem, (119)–(124), it is not necessary to require that the desired solution be bounded in the zone D. The formula (125) also gives the unique and stable solution of this problem under the condition that the desired solution $u(x,t)$ can become infinite and the function $\varphi(x)$ is given such that on the right-hand side of this formula the sub integral function and its derivatives in x and t (to its necessary order) are still absolutely and uniformly integrable functions.

The solutions of heat-conduction equations with the indicated properties are usually said to be of the well-posed class of Cauchy-Dirichlet's problem.

The following formula is evidently a multi-dimensional analog of (125)

$$\begin{aligned} u(x,t) &= \frac{1}{2^n(\pi t)^{\frac{n}{2}}} \int_{-\infty}^{\infty} \varphi(\xi) e^{-\frac{|\xi|-x^2}{4t}} d\xi , \\ d\xi &= d\xi_i \ldots d\xi_n . \end{aligned} \tag{137}$$

Chapter III. Elliptic Type Equations

§3.1 Smoothness Classes of Functions and Their Domain of Specification

3.1.1. In an Euclidean space E_n of points with Cartesian orthogonal coordinates, the distance between two points $x = (x_1, \dots, x_n)$ and $y = (y_1, \dots, y_n)$ is given by the formula

$$|x - y| = \left[\sum_{i=1}^{n} (x_i - y_i)^2\right]^{\frac{1}{2}} .$$

We take a ball $|y - x| < \varepsilon$ with center at point x and radius ε as a neighbourhood of the point x, and as usual we introduce the concepts of a limit point, an isolated point, an interior point of a set $M \in E_n$, and the concepts of a bounded set, a closed set, an open set, a compact set and the closure $\overline{M}$ of a set M. The intersection $\overline{M} \cap C M = \partial M$, where by CM we denote the complement of a set M with respect to the whole space E_n, is called the *boundary* of M.

A set M is said to be *connected* if it is impossible to find two open sets $O_1 \subset E_n$ and $O_2 \subset E_n$ such that

$$M \subseteq O_1 \cup O_2 , \quad M \cap O_1 \neq \emptyset , \quad M \cap O_2 \neq \emptyset , \quad M \cap O_1 \cap O_2 \neq \emptyset ,$$

where $\emptyset$ is an empty set.

An open connected set is called a *domain* and the closure of a domain is called a *closed domain*.

A connected subset K of a set $M \subset E_n$ with the property that there is no other connected subset K_1 of M such that $K \subset K_1$ is said to be a *component* of this set. A component of a closed set is sometimes called a *continuum*. In this definition of continuum a point can also be shown to be a continuum.

The number of components or the boundary of a domain is called the *order of connectedness* for this domain.

The concepts of continuity, uniform continuity, derivatives and differentiations of a univalent function $u(x)$ given in a domain D are supposed to have been well-known in courses of mathematical analysis.

We say that an univalent function $u(x)$ defined in a domain D is *continuous in the sense of Hölder* (or it satisfies *Hölder's condition*) if there exist positive numbers L and $h, 0 < h \leq 1$, such that for any pair of points x and y in D the estimate

$$|u(x) - u(y)| \leq L|x - y|^h \tag{1}$$

holds. If $h = 1$ in the estimate (1), we say that the function $u(x)$ is *continuous in D in Lipschitz's sense* (or it satisfies *Lipschitz's condition*).

If a function $u(x)$ in a domain D has partial derivatives of the k-th order which are continuous in Hölder's sense, then we say that $u(x)$ belongs to the class $\mathbf{C}^{k,h}(D)$. The class $\mathbf{C}^{k,0}(D)$ contains all functions given in D such that their derivatives of k-th order are continuous in D.

When a function $u(x)$, along with its derivatives up to k-th order (including itself), have limits at every point $x^0 \in \partial D$ as points x in the domain D tend to the x^0 along any path, these limits can be taken to be the values of $u(x)$ and its derivatives on the boundary ∂D of the domain D, and hence it is assumed that the function $u(x)$ and its partial derivatives, up to k-th order (including the k-th order), are defined in the closed domain $D \cup \partial D$. The meanings of notations $u(x) \in \mathbf{C}^{k,h}(D \cup \partial D)$ and $u(x) \in \mathbf{C}^{k,0}(D \cup \partial D)$ are clear.

As it is well known, a two-dimensional Euclidean space E_2 of points x with Cartesian orthogonal coordinates x_1 and x_2 can be identified with the plane of a complex variable $z = x_1 + ix_2$.

3.1.2. A set of points which satisfies the equation

$$z = z(t) = x_1(t) + ix_2(t) , \tag{2}$$

where $x_1(t)$ and $x_2(t)$ are given continuous real functions of a real parameter t on a segment

$$\alpha \leq t \leq \beta \tag{3}$$

is said to be a continuous curve on the plane E_2. As a continuous image of the compact connected set (3), the continuous curve (2) is a connected compact set.

A continuous curve is said to be a *Jordan curve* or a simple curve if, in its representation (2), different values of the parameter t correspond to different values of $z(t)$ except perhaps for $t = \alpha$ and $t = \beta$. A Jordan curve is said to be closed if $z(\alpha) = z(\beta)$. A closed Jordan curve can be considered as a topological image of a circumference.

Jordan proved that a closed simple curve divides the extended complex plane into two domains, the interior (which does not contain the point $z = \infty$) and the exterior (which does not contain the point $z = \infty$). Evidently, either of these is a simply connected domain.

When the parameter t changes in its own segment of specification (3) along a direction (from α to β or conversely), the point $z(t)$ goes around on the Jordan curve Γ. If the finite domain D^+ bounded by the closed Jordan curve still remains on the left when the going around takes place, then this direction of the going around is said to be *positive*. In the case of an open Jordan curve Γ we can also consider the direction which corresponds to the increase of the parameter t as the positive one.

A Jordan curve Γ given by equation (2) is said to be *smooth* if the functions $x_1(t)$ and $x_2(t)$ have continuous derivatives $x_1'(t)$ and $x_2'(t)$ respectively, $z'(t) = x_1'(t) + ix_2(t) \neq 0$ on the whole segment (3), and, furthermore, $z'(\alpha) = z'(\beta)$ as $z(\alpha) = z(\beta)$. At every point a smooth curve has a tangent, and the angle between it with any chosen fixed direction on the plane is a continuous function of the parameter t.

Since a smooth curve is rectifiable, it is possible to take the length s of an arc on Γ, counted from an arbitrary fixed point on Γ in the position direction, as the parameter t in the representation (2). It is clear that for the arc length element ds of the smooth curve Γ we have

$$ds = (dx_1^2 + dx_2^2)^{\frac{1}{2}} = J\ dt\ , \tag{4}$$

where

$$J^2 = ({x_1}'^2 + {x_2}'^2) \neq 0\ , \tag{5}$$

and the cosines of the outward *normals* (with respect to D^+) ν and those of the positive-direction tangents, s, to Γ are connected by the equalities

$$\cos\ \hat{\nu x_1} = \cos\ \hat{s x_2}\ , \quad \cos\ \hat{\nu x_2} = -\cos\ \hat{s x_1}\ . \tag{6}$$

Each smooth closed Jordan curve Γ has the following important property: for each Γ there exists a positive number δ_0 such that any circumsference with centre at an arbitrary $x \in \Gamma$ and radius $\delta < \delta_0$ intersects the curve Γ exactly twice.

A Jordan curve represented by the formula (2) is called *piecewise smooth* if the segment (3) can be divided into a finite set of segments so that in the interior of each one $z'(t)$ is continuous and different from zero, and at the partition points t_k the function $z'(t)$ has non-zero limits from the right side as well as from the left side.

A smooth curve Γ is called a *Lyapnov curve* if the angle $\theta(t)$ between the tangent to Γ at the point $z(t)$ and some fixed direction (for example, with the direction of the real axis in the plane of the complex variable z) is continuous in Hölder's sense.

Below we shall consider on the whole the domains D which have finite order connectedness such that every of its boundary component is either a smooth or a piecewise smooth curve, or a Lyapnov curve.

3.1.3. A surface S in the space E_3 is called a *smooth surface* if (a) it has a tangent plane everywhere which continuously varies from point to point, (b) there is a positive number δ_0 such that the part σ of S which lies in the interior of a ball with center at any point $x \in S$ and radius $\delta < \delta_0$ cuts every straight line paralleling the normal ν at point x not more than once.

Let x and y be an arbitrary pair of points in the smooth surface S, ν_x and ν_y be the unit normals to S at these points and θ be the smallest angle between ν_x and ν_y. A surface S is said to be a *Lyapnov surface* if, in addition to conditions (a) and (b), the condition (c)

$$\theta \leq L|x-y|^h$$

is satisfied, where L, h are positive numbers and $0 \leq h \leq 1$.

Taking the normal ν_x at a point $x \in S$ as the axis $O\zeta$ and selecting the axes $O\xi$ and $O\eta$ in the tangent plane to S at the point x, the equation for a part of σ of the smooth surface S can be written in the form

$$\zeta = f(\xi, \eta) \tag{7}$$

by virtue of the condition (b), where $f(\xi, \eta)$ is a real univalent function.

The condition (a) then means that the partial derivatives $\frac{\partial f}{\partial \xi}$ and $\frac{\partial f}{\partial \eta}$ exist and are continuous. Moreover

$$J = \left[1 + \left(\frac{\partial f}{\partial \xi}\right)^2 + \left(\frac{\partial f}{\partial \eta}\right)^2\right]^{-\frac{1}{2}} \neq 0\,. \tag{5$_1$}$$

In addition, for the cosines of the normal ν to S we have

$$\cos\,\hat{\nu\xi} = -\frac{1}{J}\frac{\partial f}{\partial \xi}\,, \quad \cos\,\hat{\nu\eta} = -\frac{1}{J}\frac{\partial f}{\partial \eta}\,, \quad \cos\,\hat{\nu\zeta} = \frac{1}{J}\,, \tag{6_1}$$

and for the area element ds of σ,

$$ds = J\; d\xi d\eta\;. \tag{4_1}$$

It is clear that the representation (7) of the equation for σ is equivalent to the parametric representation

$$\xi = \xi\;, \quad \eta = \eta\;, \quad \zeta = f(\xi,\eta)\;. \tag{2_1}$$

A Lyapnov surface is said to be closed if it is homomorphic to a ball.

3.1.4. We denote by S the boundary ∂D of a domain $D \in E_n$. One says that a domain D belongs to the class $A^{k,h}$ if the following conditions are fulfilled:

(1) the set S can be covered with a finite number of n-dimensional balls in each of which the moving coordinates of the points $x \in S$ allow a parametric representation

$$x_i = x_i(t_1,\dots,t_{n-1})\;, \quad i = 1,\dots,n\;, \tag{8}$$

where the functions x_i are defined and continuous in a bounded closed domain $\delta \cup \partial\delta$ in the space of variables $t_1,\dots,t_{n-1}$;

(2) for the system of functions (8), there is a one-to-one correspondence between $\delta\cup\partial\delta$ and the corresponding part of S, and all $x_i \in \mathbf{C}^{k,h}(\delta\cup\partial\delta)$, $k \geq 1$;

(3) on the whole $\delta \cup \partial\delta$, the expression

$$J = \left\{\sum_{i=1}^{n}\left[\frac{\partial(x_{i+1},\dots,x_n,x_1,\dots,x_{i-1})}{\partial(t_1,\dots,t_{n-1})}\right]^2\right\}^{\frac{1}{2}} \neq 0\;; \tag{5_2}$$

(4) the parameter representation (8) is chosen such that the cosines of an outward normal ν to S in the domain δ are given by the formula

$$\cos\,\hat{\nu x_i} = \frac{1}{J}\frac{\partial(x_{i+1},\dots,x_n,x_1,\dots,x_{i-1})}{\partial(t_1,\dots,t_{n-1})}\,, \quad i = 1,\dots,n\;. \tag{6_2}$$

The area element ds of S is written in the form

$$ds = J\ dt_1 \dots dt_{n-1} \tag{4_2}$$

using the usual notations.

A boundary S of a domain $D \subset E_n$ in the class $A^{1,h}$ is obviously an $(n-1)$-dimensional manifold which is a Lyapnov curve for $n = 2$ or a Lyapnov surface for $n = 3$. [Compare formulae $(2), (2_1), (8), (4), (4_1), (4_2), (5), (5_1), (5_2), (6)$, (6_1) and (6_2).]

§3.2 Adjoint Linear Second Order Partial Differential Operators

3.2.1. A linear second order partial differential operator has the form

$$Lu \equiv \sum_{i,j=1}^{n} A_{ij}(x)\frac{\partial^2 u}{\partial x_i \partial x_j} + \sum_{i=1}^{n} B_i(x)\frac{\partial u}{\partial x_i} + C(x)u\ , \quad A_{ij} = A_{ji}\ ,$$

where A_{ij}, B_i, C are real functions given on some domain $D_0 \subset E_n$ (cf. formula (3) of Chapter 1 and formula (4) of Chapter 2).

If $A_{ij} \in \mathbf{C}^{1,0}(D_0)$, the operator L can be rewritten in the form

$$Lu \equiv \sum_{i,j=1}^{n} \frac{\partial}{\partial x_j}\left(A_{ij}\frac{\partial u}{\partial x_i}\right) + \sum_{i=1}^{n} e_i(x)\frac{\partial u}{\partial x_i} + C(x)u\ , \tag{9}$$

where

$$e_i(x) = B_i(x) - \sum_{j=1}^{n} \frac{\partial A_{ij}(x)}{\partial x_i}\ , \quad i = 1, \dots, n\ . \tag{10}$$

In the case where $e_i(x) \in \mathbf{C}^{1,0}(D_0)$ the concept of *adjoint operator*,

$$L^* \nu \equiv \sum_{i,j=1}^{n} \frac{\partial}{\partial x_j}\left(A_{ij}\frac{\partial v}{\partial x_i}\right) - \sum_{i=1}^{n} \frac{\partial}{\partial x_i}(e_i v) + cv\ , \tag{11}$$

is introduced.

The linear operator given by formula (9) is called *selfadjoint* if the equality $Lu = L^* u$ is satisfied identically for all functions $u(x) \in \mathbf{C}^{2,0}(D)$. It is evident by (9), (10), (11) that the fulfillment of equalities

$$e_i(x) = 0\ , \quad x \in D_0\ , \quad i = 1, \dots, n\ , \tag{12}$$

is a sufficient condition for the selfadjointness of the operator L.

It is easy to see that the condition (12) is also a necessary condition for the selfadjoint operator L. In fact, applying the identity $Lu = L^*u$ to $u = 1$ and $u = x_j$ we find

$$\sum_{i=1}^{n} \frac{\partial e_i}{\partial x_i} = 0 , \quad 2e_j + x_j \sum_{i=1}^{n} \frac{\partial e_i}{\partial x_i} = 0 , \quad j = 1, \dots , n .$$

It follows from here that the equality (12) is fulfilled.

The condition (12) for the selfadjointness of the operator L can be represented in the form

$$\sum_{j=1}^{n} \frac{\partial A_{ij}(x)}{\partial x_j} = B_i(x) , \quad i = 1, \dots , n ,$$

by virtue of (10).

3.2.2. According to the definitions of operators L and L^* given by formulae (9) and (11), the identity

$$vLu - uL^*v = \sum_{i,j=1}^{n} \frac{\partial}{\partial x_j} \left[A_{ij} \left(\frac{\partial u}{\partial x_i} v - \frac{\partial v}{\partial x_i} u \right) \right] + \sum_{i=1}^{n} \frac{\partial}{\partial x_i} (e_i uv) \tag{13}$$

holds in the whole domain D_0 of these operators for any pair of functions $u(x), v(x) \in \mathbf{C}^{2,0}(D_0)$.

Let D be a subdomain in the class $A^{1,h}$ of a domain D_0. For a set of functions $p_1(x), \dots , p_n(x)$ in $\mathbf{C}^{1,0}(D \cup S)$ the Gauss-Ostrogradskii formula

$$\int_D \sum_{i=1}^{n} \frac{\partial p_i}{\partial x_i} d\tau_x = \int_s \sum_{i=1}^{n} p_i(y) \cos \, \hat{\nu y_i} ds_y \tag{G-O}$$

holds, where ν is the unit outward normal vector to the boundary $\partial D = S$ of the domain D at a point $y \in S$. We had already made use of it more than once in previous chapters.

Integrating identity (13) over the domain D and making use of formula (G-O), we obtain *Green's formula*

$$\int_D (vLu - uL^*v) d\tau_x = \int_s \left[a \left(v \frac{du}{dN} - u \frac{dv}{dN} \right) + buv \right] dS_y , \tag{14}$$

where N is the unit conormal vector at point $y \in S$ with direction cosines

$$\cos \widehat{Ny_i} = \frac{1}{a} \sum_{j=1}^{n} A_{ij} \cos \widehat{\nu y_j} \ , \quad i = 1, \ldots , n \ , \tag{15}$$

$$a^2 = \sum_{i=1}^{n} \left(\sum_{j=1}^{n} A_{ij} \cos \widehat{\nu y_j} \right)^2 ,$$

$$b = \sum_{i=1}^{n} e_i \cos \widehat{\nu y_i} \ . \tag{16}$$

Below suppose that the operator L is elliptic, i.e. its corresponding characteristic quadratic form

$$Q(\lambda_1, \ldots , \lambda_n) = \sum_{i,j=1}^{n} A_{ij} \lambda_i \lambda_j \tag{17}$$

is positively definite in the whole D_0. From this condition it follows that $a^2 > 0$. Actually since the functions A_{ij} are real, by virtue of (16) the equality $a^2 = 0$ would have meant that

$$\sum_{j=1}^{n} A_{ij} \cos \widehat{\nu y_j} = 0 \ , \quad i = 1, \ldots , n \ . \tag{18}$$

Because the determinant $A = \det \|A_{ij}\|$ of the linear homogeneous algebraic system (18) in $\cos \widehat{\nu y_j}, j = 1, \ldots , n$, is different from zero, due to the positive definiteness of the formula (17) we obtained from (b)

$$\cos \widehat{\nu y_j} = 0 \ , \quad y \in S, j = 1, \ldots , n \ .$$

But this contradicts the requirement that D be a domain in the class $A^{1,h}$.

For the scalar product $N \cdot \nu$ defined by the formula (15) of the conormal N and the normal ν at the point $y \in S$, we have

$$N \cdot \nu = \frac{1}{a} \sum_{i,j=1}^{n} A_{ij} \cos \widehat{\nu y_i} \cos \widehat{\nu y_j} \ .$$

From this we find

$$N \cdot \nu > 0 \;, \tag{19}$$

due to the positive definiteness of the form (17).

The equality (19) means that the conormal N at any point on the boundary S of a domain D in the class $A^{1,h}$ can never be a tangent plane of S.

§3.3 Existence of Solutions for Second Order Linear Elliptic Equations. Elementary Solutions.

3.3.1. Let us consider a second order linear partial differential equation.

$$Lu = f(x) \tag{20}$$

in a domain $D_0 \subset E_n$, where L is the differential operator introduced in the previous section satisfying in D_0 all the conditions of uniform ellipticity, and $f(x)$ is a given function.

By $a_{ij}(x)$ we denote the ratio of the algebraic cofactor of the element A_{ij} in matrix $\|A_{ij}\|$ to the determinant $A = \det \|A_{ij}\|$, and introduce a function $\sigma(x,y)$ of two points $x, y \in D_0$ defined by

$$\sigma(x,y) = \sum_{i,j} a_{ij}(x)(x_i - y_i)(x_j - y_j) \; . \tag{21}$$

Since the form (17) is positively definite so that the quadratic form (21) also has these properties.

Let D be a subdomain of D_0 which is the domain of specification for equation (20). Because equation (20) is uniformly elliptic in the domain D, there exist two positive numbers k_0 and k_1 such that

$$k_0|x-y|^2 \leq \sigma(x,y) \leq k_1|x-y|^2$$

on the whole $D \cup S$, where $S = \partial D$.

With respect to the coefficients of equation (20) we require that they pertained to some classes of smoothness:

$$A_{ij} \in \mathbf{C}^{3,0}(D \cup S) \;, \quad B_i, c, f \in \mathbf{C}^{1,0}(D \cup S) \;,$$

and then consider the function

$$\psi(x,y)=\begin{cases}\sigma_0(y)\sigma^{\frac{2-n}{2}}\,, & n>2\,,\\ -\dfrac{1}{4\pi\sqrt{A(y_0)}}\,, & n=2\,,\end{cases}\tag{22}$$

where

$$\sigma_0(y)=[\omega_n(n-2)\sqrt{A(y)}]^{-1}\tag{23}$$

and ω_n is the area of the unit sphere in E_n. The function $\psi(x,y)$ is called a *parametrix* or *Levi function*. If $n=2$, without loss of generality we can assume that $a_{ij}=0$ for $i\neq j$ and $a_{ij}=1$ for $i=1,2$, that is, $A(y)=1$. In the case of $n>2$, the lower bound $A(y)$ on $D\cup S$ is also positive due to the uniform ellipticity of the operator L.

Whenever $A_{ij}=0$ for $i\neq j$ and $A_{ij}=1$ for $i=1,\ldots,n$, the function defined by formula (21) is $\sigma(x,y)=|x-y|^2, A(y)=1, \sigma_0=\frac{1}{\omega_n(n-2)}$, and hence the function $\omega_n\psi(x,y)$ is just the elementary solution of Laplace's equation on the space E_n:

$$\omega_n\psi(x,y)=E(x,y)=\begin{cases}\dfrac{1}{n-2}|x-y|^{2-n}\,, & n>2\,,\\ -\log|x-y|\,, & n=2\,,\end{cases}\tag{24}$$

(cf. formula (86) of Chapter 2).

3.3.2. The function

$$u(x)=\frac{1}{\omega_n}\int_D E(x,y)\mu(y)d\tau_y\tag{25}$$

is called *Newtonian potential of a volume mass* distribute over a domain D with density μ if the integral on the right hand side is convergent.

Below it is supposed that D is a bounded domain in the class $A^{1,h}$ and $\mu\in \mathbf{C}^{1,0}(D\cup S)$.

As it is well known, the function $u(x)$ defined by the formula (25) has first order continuous derivatives on the whole E_n, and, moreover,

$$\frac{\partial u}{\partial u_i}=\frac{1}{\omega_n}\int_D\frac{\partial}{\partial x_i}E(x,y)\mu(y)d\tau_y\,,\quad i=1,\ldots,n\,.\tag{26}$$

This function is clearly harmonic outside the closed domain $D \cup S$, and, as $|x| \to \infty$, it vanishes for $n > 2$ or behaves like

$$-\log|x| \int_D \mu(y) d\tau_y$$

for $n = 2$.

Now we show that the function $u(x)$ in the domain D also has second order derivatives.

In fact, on account of the equalities

$$\frac{\partial E(x,y)}{\partial x_i} = -\frac{\partial E(x,y)}{\partial y_i}, \quad i = 1, \ldots, n,$$

we can rewrite formula (26) in the form

$$\frac{\partial u}{\partial x_i} = -\frac{1}{\omega_n} \int_D \mu(y) \frac{\partial}{\partial y_i} E(x,y) d\tau_y ,$$

or, after integrating by parts,

$$\frac{\partial u}{\partial x_i} = -\frac{1}{\omega_n} \int_s E(x,y)\mu(y) \cos \hat{\nu y_i} ds_y + \frac{1}{\omega_n} \int_D \frac{\partial u}{\partial y_i} E(x,y) d\tau_y .$$

From this follows the existence of the second order derivatives of the function $u(x)$ at every point $x \in D$ and, moreover,

$$\begin{aligned} \frac{\partial^2 u}{\partial x_i \partial x_j} &= -\frac{1}{\omega_n} \int_s \frac{\partial E(x,y)}{\partial x_j} \mu(y) \cos \hat{\nu y_i} ds_y + \frac{1}{\omega_n} \int_D \frac{\partial \mu}{\partial y_i} \frac{\partial E(x,y)}{\partial x_j} d\tau_y \\ &= \frac{1}{\omega_n} \int_s \frac{\partial E(x,y)}{\partial y_j} \mu(y) \cos \hat{\nu y_i} ds_y - \frac{1}{\omega_n} \int_D \frac{\partial \mu}{\partial y_i} \frac{\partial E(x,y)}{\partial y_j} d\tau_y . \end{aligned} \tag{27}$$

By (27) we have

$$\Delta u = \frac{1}{\omega_n} \int_s \frac{\partial E(x,y)}{\partial \nu_y} \mu(y) ds_y - \lim_{\varepsilon \to 0} \frac{1}{\omega_n} \int_D \sum_{i=1}^n \frac{\partial \mu}{\partial y_i} \frac{\partial E(x,y)}{\partial y_i} d\tau_y , \tag{28}$$

where D_ε is the part of the domain D that is exterior to the closed ball $|y-x| \leq \varepsilon$ and inside D.

Because of

$$\sum_{i=1}^{n} \frac{\partial^2 E(x,y)}{\partial y_i^2} = 0$$

for $x \neq y$, and thus

$$\sum_{i=1}^{n} \frac{\partial \mu}{\partial y_i} \frac{\partial E}{\partial y_i} = \sum_{i=1}^{n} \frac{\partial}{\partial y_i} \left(\mu \frac{\partial E}{\partial y_i} \right) ,$$

by integrating by parts and applying the formula (G-O) we can write

$$\begin{aligned} \int_{D_\varepsilon} \sum_{i=1}^{n} \frac{\partial \mu}{\partial y_i} \frac{\partial E(x,y)}{\partial y_i} d\tau_y &= \int_{D_\varepsilon} \sum_{i=1}^{n} \frac{\partial}{\partial y_i} \left(\mu \frac{\partial E(x,y)}{\partial y_i} \right) d\tau_y \\ &= \int_s \mu(y) \frac{\partial E(x,y)}{\partial \nu_y} ds_y - \int_{|y-x|=\varepsilon} \frac{\partial E(x,y)}{\partial \nu_y} ds_y . \end{aligned} \tag{29}$$

On the basis of (28) and (29) for $x \in D$,

$$\begin{aligned} \Delta u &= \frac{1}{\omega_n} \lim_{\varepsilon \to 0} \int_{|y-x|=\varepsilon} \mu(y) \frac{\partial E(x,y)}{\partial \nu_y} ds_y \\ &= -\frac{1}{\omega_n} \lim_{\varepsilon \to 0} \int_{|y-x|=\varepsilon} \mu(y) \frac{ds_y}{|y-x|^{n-1}} \\ &= -\frac{1}{\omega_n} \mu(x) \lim_{\varepsilon \to 0} \int_{|y-x|=\varepsilon} \frac{ds_y}{\varepsilon^{n-1}} - \frac{1}{\omega_n} \lim_{\varepsilon \to 0} \int_{|y-x|=\varepsilon} \frac{\mu(y)-\mu(x)}{|y-x|^{n-1}} ds_y \\ &= -\mu(x) . \end{aligned} \tag{30}$$

The requirement that D is a domain in the class $A^{1,h}$ can be avoided when deriving formula (30). In fact, we can rewrite the expression (25) of the function $u(x)$ in the form

$$u(x) = \frac{1}{\omega_n} \int_{D_R} E(x,y)\mu(y) d\tau_y + \frac{1}{\omega_n} \int_{|y-x|<R} E(x,y)\mu(y) d\tau_y ,$$

where the ball $|y-x| \leq R$ lies in the domain D, and D_R is the part of D that is outside the ball $|y-x| \leq R$. On the right side of the representation, the first term is a harmonic function in the ball $|y-x| < R$ and the second term is clearly suitable for the argument mentioned above.

The equality (30) means that the Newtonian potential (25) of a mass distributed over the domain D with density μ is a regular solution for the Poisson equation

$$\Delta\mu = -\mu(x) . \tag{31}$$

3.3.3. As already noticed before, in the case of two independent variables we can suppose that the principle part of the elliptic operator L coincides with the Laplacian and equation (20) is then written in the form

$$\Delta u + \sum_{i=1}^{2} B_i(x)\frac{\partial u}{\partial x_i} + c(x)u = f(x) , \quad x = (x_1, x_2) . \tag{32}$$

Below we look for a solution $u(x)$ of equation (32) according to the formula $u(x) = \omega(x)+v(x)$, where $\omega(x)$ is an arbitrary function in the class $\mathrm{C}^{3,0}(D \cup S)$ and $v(x)$ is the potential of mass with density $\mu(x)$ distributed over a bounded domain D,

$$v(x) = -\frac{1}{2\pi}\int_D \log|x-y|\mu(y)d\tau_y . \tag{33}$$

Taking into account that the function $v(x)$ defined by (33) is a solution of the Poisson equation $\Delta v = -\mu(x), x \in D$, we arrive at the conclusion that the function $u(x)$ will be a solution of equation (32) if

$$\mu(x) + \int_D K(x,y)\mu(y)d\tau_y = g(x) , \tag{34}$$

where

$$K(x,y) = \frac{1}{2\pi}\left[\sum_{i=1}^{2} B_i(x)\frac{x_i - y_i}{|x-y|^2} + c(x)\log|x-y|\right] \tag{35}$$

and

$$g(x) = L\omega(x) - f(x) . \tag{36}$$

The equality (34) is a second kind Fredholm's integral equation with respect to μ since its kernel $K(x,y)$ has a singularity of the form

$$K(x,y) = O\left(\frac{1}{|x-y|}\right) .$$

As it is well known, a second kind Fredholm's integral equation always has solutions in a domain of sufficiently small area (measure). Therefore, the equation (34) in a domain D of sufficiently small area has a solution $\mu(x)$ which belongs to the class $\mathbf{C}^{1,0}(D \cup S)$ under the above assumptions with respect to B_i, c and f on account of (36). Hence the formula $u(x) = \omega(x) + v(x)$ gives a family of regular solutions for equation (32) in the domain D, which depends on the arbitrary function $\omega(x) \in \mathbf{C}^{3,0}(D \cup S)$.

Consider the corresponding to (32), homogeneous equation

$$Lu = 0 \tag{37}$$

with coefficients B_i, C in the class $\mathbf{C}^{1,0}(D \cup S)$ and introduce for discussion a function $\Omega_0(x,y)$ of two points $x, y \in D$ by using the formula

$$\Omega_0(x,y) = -\frac{1}{2\pi}\log|x-y| - \frac{1}{2\pi}\int_D \log|x-t|\mu(t)d\tau_t \ , \tag{38}$$

where μ is a so far unknown function.

As a function of the point x the expression (38) for $x \neq y$ will be the solution of equation (37) as long as the equality

$$\mu(x) + \int_D K(x,t)\mu(t)d\tau_t = g(x,y) \tag{39}$$

holds, where $K(x,t)$ is given by (35) if $y = t$ in it and

$$g = -\frac{1}{2\pi}L(\log|x-y|) \ .$$

Since as $|x-t| \to 0$ and $|x-y| \to 0$ the estimates

$$K(x,t) = O\left(\frac{1}{|x-t|}\right) \quad \text{and} \quad g(x,y) = O\left(\frac{1}{|x-y|}\right)$$

hold, with respect to μ the equality (39) is a second kind Fredholm's integral equation which always has at least a solution in a domain D of sufficiently small area (measure). Evidently, a solution μ of this equation is a function of the points x and y, and the estimate

$$\mu(x,y) = O\left(\frac{1}{|x-y|}\right)$$

holds for $|x-y| \to 0$.

Substituting this $\mu(x,y)$ into the right-hand side of formula (38) we obtain the function

$$-\frac{1}{2\pi}\log\,|x-y|-\frac{1}{2\pi}\int_D \log|x-t|\mu(t,y)d\tau_t\ ,$$

which has the properties: (1) in the domain D for $x \neq y$ it is a regular solution in x of equation (37); (2) near the point y as $|x-y| \to 0$ the estimates

$$\Omega_0 = \left(\log\frac{2R}{|x-y|}\right) \quad \text{and} \quad \frac{\partial_0\Omega}{\partial x_i} = O\left(\frac{1}{|x-y|}\right)$$

hold, where $2R$ is the diameter of the domain D.

Functions with properties (1) and (2) are called *elementary solutions* of the homogeneous equation (37). Under the assumptions concerning coefficients B_i and C used, the above equation (37) in any case always has an elementary solution in a domain of sufficiently small area (measure).

3.3.4. In the theory of elliptic equations, elementary solutions play an essential role. Therefore, the way of constructing them is one of the most important problems for this theory.

A function $\Omega_0(x,y)$ of two points x and y in a domain $D \subset E_n$ is called an elementary solution of the homogeneous equation, corresponding to (20),

$$Lu \equiv \sum_{i,j=1}^{n} A_{ij}\frac{\partial^2 u}{\partial x_i \partial x_j} + \sum_{i=1}^{n} B_i \frac{\partial u}{\partial x_i} + Cu = 0 \tag{40}$$

with $n > 2$, if it has the properties: (1) when $x \neq y$ the function $\Omega_0(x,y)$ in x is a regular solution of equation (40); (2) the estimates

$$\Omega_0 = O(|x-y|^{2-n}) \quad \text{and} \quad \frac{\partial\Omega_0}{\partial x_i} = O(|x-y|^{1-n})\ , i = 1, \ldots, n,$$

hold as $|x-y| \to 0$.

Under the assumptions that $A_{ij} \in \mathbf{C}^{3,0}(D \cup S)$ and $B_i, C \in \mathbf{C}^{1,0}(D \cup S)$, the method stated above for constructing an elementary solution is also successfully applied to the case of equation (40) with n (the number of independent variables) > 2, as long as the operator L is uniformly elliptic.

In fact, for the first and second order derivatives of the Levi function defined by the first of formulas (22),

$$\psi(x,y) = \sigma_0(x,y)\sigma^{\frac{2-n}{2}} ,$$

we have by (21)

$$\frac{\partial\psi}{\partial x_i} = -\sigma_0(y)(n-2)\sigma^{-\frac{n}{2}}\left[\sum_{k=1}^{n} a_{ik}(x_k-y_k) + \frac{1}{2}\sum_{l,k=1}^{n}\frac{\partial a_{lk}}{\partial x_i}(x_l-y_l)(x_k-y_k)\right.$$

$$- \sigma_0(y)(n-2)\sigma^{-\frac{n}{2}}\sum_{k=1}^{n} a_{ik}(x_k-y_k) + P_i(x,y) , \tag{41}$$

$$\frac{\partial^2\psi}{\partial x_i\partial x_y} = -\sigma_0(y)(n-2)\sigma^{-\frac{n+2}{2}}\left[a_{ij}\sigma - n\sum_{k,l=1}^{n} a_{ik}a_{jl}(x_l-y_l)(x_k-y_k)\right]$$

$$+ P_{ij}(x,y) , \tag{42}$$

$i,j = 1,\dots,n.$
where for $\frac{\partial\psi}{\partial x_i}$, $P_i(x,y)$ and $P_{ij}(x,y)$ we have the estimates

$$\frac{\partial\psi}{\partial x_i} = O(|x-y|^{1-n}) , \quad P_i = O(|x-y|^{2-n}) , \quad P_{ij} = O(|x-y|^{1-n}) \tag{43}$$

as $|x-y| \to 0$.

Since

$$\sum_{i=1}^{n} A_{ij}a_{ik} = \begin{cases} 0 , & k \neq j \\ 1 , & k = j \end{cases}$$

from (42) and taking into account (21) we find

$$\sum_{i,j=1}^{n} A_{ij}\frac{\partial^2\psi}{\partial x_i\partial_j} = -\sigma_0(y)(n-2)\sigma^{\frac{n+2}{2}}\left[n\sigma - n\sum_{k,l=1}^{n} a_{kl}(x_k-y_k)(x_l-y_l)\right]$$

$$+ \sum_{i,j=1}^{n} A_{ij}P_{ij} = \sum_{i,j=1}^{n} A_{ij}P_{ij} . \tag{44}$$

On the basis of (41), (43) and (44) we conclude that as $|x-y| \to 0$ the estimate

$$L\psi(x,y) = 0(|x-y|^{1-n}) \tag{45}$$

holds, where the operator L is taken with respect to x.

Now let us consider a function

$$v(x) = \int_D \psi(x,y)\mu(y)d\tau_y \ , \tag{46}$$

where $\mu(x) \in \mathbf{C}^{1,0}(D \cup S)$ and D is a bounded domain in the class $A^{1,h}$.

As in the case of Newtonian potential (25) it is not difficult to show that for $x \in D$

$$Lv = -\mu(x) + \int_D L\psi(x,y)\mu(y)d\tau_y \ . \tag{47}$$

Due to (45) the integral on the right-hand side of formula (47) converges absolutely and uniformly

Repeating the above argument for the case $n = 2$ we conclude that if the function $\mu(x)$ is a solution of the integral equation

$$\mu(x) + \int_D K(x,y)\mu(y)d\tau_y = -(f - L\omega) \ , \tag{48}$$

where $K(x,y) = -L\psi$, then the function $u(x)$ defined by

$$u(x) = \omega(x) + \int_D \psi(x,y)\mu(y)d\tau_y \tag{49}$$

is a regular solution in the domain D of the non-homogeneous equation (20) with coefficients $A_{ij} \in \mathbf{C}^{3,0}(D \cup S)$ and $B_i, c, f \in \mathbf{C}^{1,0}(D \cup S)$ provided the arbitrary function $\omega(x) \in \mathbf{C}^{3,0}(D \cup S)$.

By virtue of the estimate (45), the equality (48) in $\mu(x)$ is a second kind Fredholm's integral equation which always has a solution in the space of functions $\mathbf{C}^{1,0}(D \cup S)$, at least if the domain D is of a sufficiently small area (measure). Thus, in the domains of sufficiently small measure the non-homogeneous equation (20) has a family of regular solutions which depends on an arbitrary function $\omega(x)$.

If a solution $\Omega_0(x, y)$ of the homogeneous equation (40) may be sought in the form

$$\Omega_0(x,y) = \psi(x,y) + \int_D \psi(x,t)\mu(t)d\tau_t \ , \tag{50}$$

then, instead of (48) for the unknown density $\mu(x)$, we will obtain a second kind Fredholm's integral equation

$$\mu(x) + \int_D K(x,t)\mu(t)d\tau_t = L\psi \ ,$$

where the kernel is the same as the one of equation (48). Thus, we conclude that the homogeneous equation (40) in a domain of sufficiently small measure has a solution of the form (50) which satisfies all the conditions above produced for elementary solutions.

The function $u_0(x)$ defined by

$$u_0(x) = \int_D \Omega_0(x,y)\mu(y)d\tau_y \tag{51}$$

is naturally called a *generalized potential of mass* with a density $\mu(x)$ distributed over the domain D. It is clear that when $\mu \in \mathbf{C}^{1,0}(D \cup S)$ the function $u_0(x)$ is a regular solution in the domain D of the non-homogeneous equation

$$Lu_0(x) = -\mu(x) \ .$$

The stated above property (2) of the elementary solution $\Omega_0(x, y)$ for $n > 2$ is required to be more precise with respect to its structure if $x = y$. Such precision becomes possible if the coefficients in equation (40) are analytic functions. Under this assumption, denoted by Γ the square of the *geodesic distance* with the points x and y in a Riemann space with a metric $ds^2 = \sum_{i,j=1}^n a_{ij} dx_i dx_j$, we can claim that for odd n

$$\Omega_0(x,y) = R_1 \Gamma^{\frac{2-n}{n}} + R_2 \ ,$$

and for even n

$$\Omega_0(x,y) = R_1 \Gamma^{\frac{2-n}{n}} + R_3 \log \ \Gamma + R_2 \ ,$$

where R_1, R_2 and R_3 are analytic functions of variables $x_1, \ldots, x_n$, and moreover, it may happen that R^3 vanishes identically.

The existence of an elementary solution $\Omega_0(x,y)$ for equation (40) has been proved above for a domain of sufficiently small measure, or, as we usually say, in a small domain.

3.3.5. Now we suppose that equation (40) has been given on the whole space E_n. An elementary solution $\Omega(x,y)$ defined on E_n of equation (40) will be called a *principle elementary solution* if there exists a constant $\alpha > 0$ such that

$$\Omega = O(e^{-\alpha|x-y|}), \quad \frac{\partial \Omega}{\partial x_i} = O(e^{-\alpha|x-y|}), \quad (i = 1, \ldots, n) \tag{52}$$

as $|x-y| \to \infty$.

We mention without proof the following important assertion: the elliptic equation (40) given on the whole space E_n has a principal elementary solution $\Omega(x,y)$ if (1) the function $A(x) = \det\|A_{ij}\|$ is bounded below by a positive number; (2) the functions A_{ij}, B_i and C are bounded and continuous in Hölder's sense, and moreover, A_{ij} are uniformly continuous in Hölder's sense; (3) the function $c(x) \leq 0$ in the whole E_n and outside some bounded domain $c(x) < -g^2$, where g is a nonzero real constant. Besides, when it is known that $A_{ij}, e_i, i, j = 1, \ldots, n$ have first order derivatives which are bounded and uniformly continuous in Hölder's sense in the whole E_n, then for equation (40) there is a principal elementary solution $\Omega(x,y)$ which satisfies the adjoint homogeneous equation

$$L^*\Omega = 0 \tag{53}$$

with respect to y, as $y \neq x$.

Where the equation (40) which is given in a bounded domain $D \cup E_n$ is uniformly elliptic, its coefficients are sufficiently smooth functions, and $c(x) \leq 0$, then this equation can be determined on the whole space E_n such that the conditions (1), (2) and (3) are satisfied. Therefore in the studied case we suppose at first that equation (40) has a principal elementary solution.

It can be verified by a direct check that the function

$$\Omega(x,y) = \begin{cases} \varphi_2(r), & n = 2 \\ \varphi_3(r), & n = 3, \end{cases}$$

where

$$\varphi_2(r) = \frac{1}{2\pi}\int_1^\infty \frac{1}{\sqrt{t^2-1}}e^{-rt}dt\ , \quad \varphi_3(r) = \frac{1}{4\pi r}e^{-r}\ , \quad r = |x-y|\ ,$$

is a principal elementary solution of the equation

$$\Delta u - u = 0\ , \tag{$*$}$$

for the cases of two and three independent variables.

It should be noticed that if $\varphi_n(r)$ is a principal elementary solution of the equation ($*$) with n, then

$$\varphi_{n+2} = -\frac{1}{2\pi r}\frac{d}{dr}\varphi_n(r)\ ,$$

where $\varphi_n(r)$ is the solution of the ordinary differential equation

$$r\varphi''(r) + (n-1)\varphi'(r) - r\varphi(r) = 0\ .$$

It is easy to see that the function $\Omega(x,y) = \lambda^{n-2}\varphi_n(\lambda r)$ is a principal elementary solution for the equation

$$\Delta u - \lambda^2 u = 0\ , \quad \lambda = \text{constant}\ .$$

An equation in the form of (40) which is elliptic on the whole space E_n with analytic coefficients always (i.e. without the requirement $c \le 0$) has an elementary solution that must not possess the property (52) at all.

§3.4 Double Layer and Simple Layer Potentials

3.4.1. Let D be a bounded domain in E_n with a smooth boundary $S = \partial D, E(x,y)$ be the elementary solution of Laplace's equation and $\mu(y)$ be an integrable function given on S.

The functions $v(x)$ and $w(x)$ defined by the formulae

$$v(x) = \frac{1}{\omega_n}\int_S \frac{\partial E(x,y)}{\partial \nu_y}\mu(y)ds_y\ , \tag{54}$$

$$w(x) = \frac{1}{\omega_n}\int_S E(x,y)\mu(y)ds_y \tag{55}$$

are said to be a *double layer potential* and *simple layer potential* of mass distributed on a surface S with density μ respectively.

From the formula (87) in Chapter II it is certainly true that for any arbitrary function $u(x)$ which is harmonic in a domain D of the class $A^{1,h}$, being in the class $\mathbf{C}^{1,0}(D \cup S)$, this function can be represented at every point $x \in D$ in the form of a sum of a double layer potential and a simple layer potential of mass distributed on S with density $\mu(y) = -u(y)$ and $\mu(y) = \frac{du}{d\nu_y}$ respectively.

Just as before, for the domain D and the complement of $D \cup S$ with respect to the space E_n we shall use the notations D^+ and D^-. It follows from the formula (86) in Chapter II for the elementary solution $E(x, y)$ of Laplace's equation that $v(x)$ and $w(x)$ are harmonic in both D^+ and D^-; moreover, the function $v(x)$ vanishes for all $n \geq 2$ as $|x| \to \infty$, while the function $w(x)$ vanishes for $n > 2$ and behaves like

$$-\frac{1}{2\pi} \log\ |x| \int_S \mu(y) ds_y$$

for $n = 2$.

3.4.2. Let us turn to an investigation of the basic properties of $v(x)$ and $w(x)$. Since these properties are independent of whether $n = 2$ or $n > 2$, we confine ourselves to observing the case of $n = 2$ below and suppose that S is a closed Jordan curve with continuous curvature, i.e., D is a domain in the class $A^{2,0}$, and $\mu \in \mathbf{C}^{2,0}(S)$.

Since $n = 2$, the expression (54) for the double layer potential $v(x)$ is written in the form

$$v(x) = -\frac{1}{2\pi} \int_S \frac{\partial}{\partial \nu_y} \log |y - x| \mu(y) ds_y\ ,$$
$$x = (x_1, x_2)\ , \quad y = (y_1, y_2)\ . \tag{56}$$

The arc abscissas of the points $x^0, y \in S$ (i.e the lengths of the arcs that counted from some fixed point on S to the points x^0 and y along the counter clockwise direction) will be denoted by s and t respectively.

Now we show that the double layer potential (56) still makes sense if $x = x^0 \in S$. If fact, for the function

$$\pi k(s, t) = \frac{\partial}{\partial \nu_y} \log |y - x^0|$$

we have

$$\pi k(s,t) = \frac{1}{|y - x^0|^2} \sum_{i=1}^{2} (y_1 - x_i^0) \cos \hat{\nu y_i}$$
$$= \frac{\cos \varphi}{|y - x^0|} = \frac{\partial}{\partial t} \vartheta(s,t) , \tag{57}$$

where

$$\cos \varphi = \frac{(y - x^0) \cdot \nu_y}{|y - x^0|} , \quad \vartheta(s,t) = \operatorname{arctg} \frac{y_2 - x_2^0}{y_1 - x_1^0} .$$

It is easy to see that the function $k(s,t)$ is continuous on S in the union of variables s and t.

If fact, because the cosines of the unit outward normal $(\cos \hat{\nu y_1}, \cos \hat{\nu y_2})$ and the unit tangent vector (y_1', y_2') at the point $y \in S$ are connected to each other by the formulae

$$\cos \hat{\nu y_1} = y_2' , \quad \cos \hat{\nu y_2} = -y_i' ,$$

we can write the expression (57) for $k(s,t)$ in the form

$$k(s,t) = \frac{1}{\pi} \frac{(y_1 - x_1^0) y_2' - (y_2 - x_2^0) y_1'}{|y - x|^2} .$$

From this and considering that the equalities

$$y_1 - x_1^0 = (t - s) x_1^{0'} + \frac{(t-s)^2}{2} \tilde{y}_1'' ,$$
$$y_2 - x_2^0 = (t - s) x_2^{0'} + \frac{(t-s)^2}{2} \tilde{y}_2'' ,$$
$$y_1' - x_1^{0'} = (t - s) \tilde{y}_1'' ,$$

and

$$y_2' - x_2^{0'} = (t - s) \tilde{y}_2''$$

are true under the adopted assumptions concerning smoothness of the curve S, we find, as $t \to s$,

$$\lim_{t \to s} k(s,t) = \frac{1}{2\pi} k(s) ,$$

where $k(s)$ is the *curvature* of the curve S at the point x^0.

Defining $k(s,t)$ for $t = s$ as $\frac{1}{2\pi}k(s)$ we verified the validity of the stated assertion.

It follows from the continuity of $k(s,t)$ that the double layer potential (56)

$$v(x^0) = \frac{1}{2\pi}\int_S \frac{\cos\varphi}{|y - x^0|}\mu(y)ds_y \ , \tag{58}$$

makes sense and is a continuous function of x^0 on S.

A basic property of the double layer potential $v(x)$ can be stated in the form of the following assertion: the limits

$$\lim_{\substack{x \to x^0 \\ x \in D^+}} v(x) = v^+(x^0) \ , \quad \lim_{\substack{x \to x^0 \\ x \in D^-}} v(x) = v^-(x^0) \tag{59}$$

exist at every point $x^0 \in S$; moreover

$$v^+(x^0) - v(x^0) = -\frac{1}{2}\mu(x^0) \ , \tag{60}$$

$$v^-(x^0) - v(x^0) = \frac{1}{2}\mu(x^0) \ . \tag{61}$$

The stated assertion can be verified considerably simply under the assumptions adopted above concerning the smoothness of the curve S and the function $\mu(x)$. Actually, we denote by d the circle $|x - x^0| < \varepsilon$ with centre at the point x^0 and sufficiently small radius ε, and by S' the part of the S inside d. Let $d' = d \cap D^+$ and $v_1(x)$ be a real-valued function in the class $\mathbf{C}^{2,0}(d' \cup \partial d')$ which satisfies the boundary conditions

$$v_1(x) = \mu(x) \ , \quad \frac{\partial v_1}{\partial \nu_x} = 0 \ , \quad x \in S' \ . \tag{62}$$

We denote the part of the circumference $|x - x^0| = \varepsilon$ that is inside the D^+ by σ.

Integrating the identity

$$\sum_{i=1}^{2} \frac{\partial}{\partial y_i}\left(\log|y - x|\frac{\partial v_1}{\partial y_i} - v_1\frac{\partial}{\partial y_i}\log|y - x|\right)$$
$$= \log|y - x|\Delta v_1 - v_1\Delta\log|y - x|$$

on the domain d' (if $x \in d' \cup S'$, then extract from $d' \cup S'$ this point together with the closed circle $|y - x| \leq \sigma$ of sufficiently small radius δ, and take integration on the remaining part of the domain d' and let δ tend to zero) and taking into account into the equality (62), we find

$$-\int_S \mu \frac{\partial}{\partial \nu_y} \log|y-x| ds_y + \int_\sigma \left(\log|y-x| \frac{\partial v_1}{\partial \nu_y} - \nu_1 \frac{\partial}{\partial \nu_y} \log|y-x| \right) ds_y$$
$$+ q(x) v_1(x) = \int_{d'} \log|y-x| \Delta v_1 d\tau_y , \tag{63}$$

where

$$q(x) = \begin{cases} 2\pi , & x \in d' \\ \pi , & x \in S' \\ 0 , & x \in D^- . \end{cases} \tag{64}$$

Rewriting the formula (56) in the form

$$v(x) = -\frac{1}{2\pi} \int_{S'} \mu \frac{\partial}{\partial \nu_y} \log|y-x| ds_y - \frac{1}{2\pi} \int_{S''} \mu \frac{\partial}{\partial \mu_y} \log|y-x| ds_y ,$$

where S'' is the part of S that is outside the circle d, and making use of the equality (63), we obtain

$$v(x) = -\frac{1}{2\pi} \int_{S''} \mu \frac{\partial}{\partial \nu_y} \log|y-x| ds_y + \frac{1}{2\pi} \int_\sigma \left(v_1 \frac{\partial}{\partial \nu_y} \log|y-x| \right.$$
$$\left. - \log|y-x| \frac{\partial v_1}{\partial \nu_y} \right) ds_y + \frac{1}{2\pi} \int_{d'} \log|y-x| \Delta v_1 d\tau_y - \frac{1}{2\pi} q(x) v_1(x) . \tag{65}$$

These integrated terms in the right-hand side of the formula (65) are continuous while the point x passes through x^0 from D^+ to D^-. Taking into account this property and the equality (64) for quantities $v(x_0)$, $v^+(x^0)$ and $v^-(x^0)$ defined by the formulae (58) and (59), we obtain the relations (60) and (61). Therefore, the double layer potential has a discontinuity with jumps expressed by the formulae (60) and (61) when the point $x \to x^0$, where x is kept in D^+ or in D^- all the time.

In view of the fact that the integrated terms on the right-hand side of (65) have continuous first order derivatives when the point x passes through the point $x^0 \in S$ from D^+ and D^- due to the second equality in (62) and (64), we conclude that the limits

$$\nu_{x_0} \cdot \lim_{\substack{x \to x^0 \\ x \in D^+}} \operatorname{grad} v(x) = \left(\frac{\partial v}{\partial \nu_{x_0}} \right)^+ ,$$

$$\nu_{x_0} \cdot \lim_{\substack{x \to x^0 \\ x \in D^-}} \operatorname{grad} v(x) = \left(\frac{\partial v}{\partial \nu_{x_0}} \right)^-$$

exist and that

$$\left(\frac{\partial v}{\partial \nu_{x_0}} \right)^+ - \left(\frac{\partial \nu}{\partial \nu_{x_0}} \right)^- = 0 \tag{66}$$

at every point $x^0 \in S$.

3.4.3. We now consider the simple layer potential (55) with $n = 2$

$$w(x) = \frac{1}{2\pi} \int_S \log |y - x| \mu(y) ds_y \ . \tag{67}$$

Repeating the process made above in reducing the formula (65), except that this time, instead of $v(x)$ and $v_1(x)$ we use, the function $w(x)$ and a function $w_1(x) \in \mathbf{C}^{2,0}(d' \cup \partial d')$ which satisfies the boundary conditions

$$w_1(x) = 0 \ , \quad \frac{\partial w_1}{\partial \nu_x} = \mu(x) \ , \quad x \in S' \tag{68}$$

respectively, we obtain,

$$\begin{aligned} &\frac{1}{2\pi} \int_{S'} \log |y - x| \mu(y) ds_y + \frac{1}{2\pi} q(x) w_1(x) \\ &\quad + \frac{1}{2\pi} \int_\sigma (\log |y - x| \frac{\partial w_1}{\partial \nu_y} - w_1 \frac{\partial}{\partial \nu_y} \log |y - x|) ds_y \\ &\quad = \frac{1}{2\pi} \int_{d'} \log |y - x| \Delta w_1 d\tau \ . \end{aligned} \tag{69}$$

We can write the simple layer potential (67) in the form

$$\begin{aligned} w(x) = &-\frac{1}{2\pi} \int_{S''} \log |y - x| \mu(y) ds_y + \frac{1}{2\pi} q(x) w_1(x) \\ &+ \frac{1}{2\pi} \int_\sigma (\log |y - x| \frac{\partial w_1}{\partial \nu_y} - w_1 \frac{\partial}{\partial \nu_y} \log |y - x|) ds_y \\ &- \frac{1}{2\pi} \int_{d'} \log |y - x| \Delta w_1 d\tau_y \end{aligned} \tag{70}$$

because of the formula (69).

Because of the continuous differentiability in the whole of circle d for integrated terms on the right-hand side of the formula and taking into account (64) and (68), we conclude that the simple layer potential $w(x)$ is still continuous while the point x passes through the point $x^0 \in S$ from the domain D^+ to the domain D^-, but its normal derivative $\frac{\partial w}{\partial \nu_x}$ has a discontinuity, so that

$$\left(\frac{\partial w}{\partial \nu_{x^0}}\right)^+ - \frac{\partial w}{\partial \nu_{x_0}} = \frac{1}{2}\mu(x^0) , \tag{71}$$

$$\left(\frac{\partial w}{\partial \nu_{x^0}}\right)^- - \frac{\partial w}{\partial \nu_{x_0}} = -\frac{1}{2}\mu(x^0) , \tag{72}$$

where

$$\left(\frac{\partial w}{\partial \nu_{x^0}}\right)^+ = \nu_{x^0} \cdot \lim_{\substack{x \to x^0 \\ x \in D^+}} \operatorname{grad} w(x) , \tag{73}$$

$$\left(\frac{\partial w}{\partial \nu_{x^0}}\right)^- = \nu_{x^0} \cdot \lim_{\substack{x \to x^0 \\ x \in D^-}} \operatorname{grad} w(x) , \tag{74}$$

$$\frac{\partial w}{\partial \nu_{x^0}} = \frac{1}{2\pi} \int_S \frac{(y - x^0)\nu_{x_0}}{|y - x^0|^2} \mu(y) ds_y = \frac{1}{2} \int_S k^*(s,t)\mu(t)dt , \tag{75}$$

$$\begin{aligned} K^*(s,t) &= -\frac{1}{\pi}\frac{\partial}{\partial s} \operatorname{arctg} \frac{y_2 - x_2^0}{y_1 - x_1^0} \\ &= \frac{1}{\pi}\frac{(y - x^0)\nu_{x_0}}{|y - x^0|^2} = -\frac{1}{\pi}\frac{\partial}{\partial \nu_{x^0}} \log |y - x^0| . \end{aligned} \tag{76}$$

Supplementing the definition of $k^*(s,t)$ for $t = s$ by $\lim\limits_{t \to s} k^*(s,t)$ we verified that this function is continuous in the union of variables s, t on the boundary S.

The properties established above for double layer and simple layer potentials remain valid under more general assumptions with respect to the boundary S of the domain D^+ and the function μ, in particular, when S is a Lyapnov curve and $\mu \in \mathbf{C}^{0,0}(S)$.

§3.5 Dirichlet Problem for Harmonic Functions

3.5.1. Let D^+ be a bounded domain in E_2 such that its boundary S is a Jordan curve with continuous curvature, and $x = (x_1, x_2) \in D^+$, $x^0(x_1^0, x_2^0) \in S$, $y = (y_1, y_2) \in S$.

Here we will consider the Dirichlet problem formulated as follows: to find a function $u(x)$ which is harmonic in the domain D^+, belonging to the class $\mathbf{C}^{0,0}(D^+ \cup S)$, and satisfying the boundary condition

$$u(x^0) = \varphi(x^0) \ , \quad x^0 \in S \ , \tag{77}$$

where $\varphi(x^0)$ is a real function of the given on S class $\mathbf{C}^{0,0}(S)$.

As it has already been proved in §7, Chapter II, the uniqueness of the solution of the problem (77) is a direct consequence of the extremum principle for harmonic functions.

Now let us turn to the proof of the existence of the solution of this problem. We will look for it in the form of a double layer potential

$$u(x) = -\frac{1}{2}\int_S \frac{\partial}{\partial \nu_y} \log|y-x|\mu(y)ds_y \ , \quad x \in D^+ \ , \tag{78}$$

with a density μ in the class of $\mathbf{C}^{0,0}(S)$, unknown at the moment.

By virtue of the property of double layer potentials expressed by (60) we have

$$w^+(x^0) = -\frac{1}{2}\mu(x^0) + u(x^0) \ , \quad x^0 \in S \ , \tag{79}$$

where

$$u(x^0) = -\frac{1}{2\pi}\int_S \frac{\cos\varphi}{|y-x^0|}\mu(y)ds_y \tag{80}$$

on account of (58).

Using the notations

$$k(s,t) = \frac{1}{\pi}\frac{\cos\varphi}{|y-x^0|} \ , \quad \mu(y) = \mu(t) \ , \tag{81}$$

where s and t are the arc abscissas of points $x^0, y \in S$, we can rewrite the formula (80) in the form

$$u(x^0) = -\frac{1}{2}\int_S k(s,t)\mu(t)dt \ . \tag{82}$$

If it is required that the harmonic function $u(x)$ represented by the formula to (52) satisfy the boundary condition (77), then we come to the conclusion, on the basis of formulae (79), (80), (81) and (82), that the function $\mu(x^0) = \mu(s)$ should be a solution of the equation

$$\mu(s) + \int_S k(s,t)\mu(t)dt = -2\varphi(s) , \tag{83}$$

where $\varphi(s) \equiv \varphi(x^0)$.

Since S is a curve with continuous curvature, the function $k(s,t)$ defined by the formula (81), as already been proved in the preceding paragraph, is continuous on S in the union of variables s, t. Therefore, the equality (83) with respect to μ is a Fredholm's integral equation of the second kind. It is clear that if μ is a solution of equation (83) then the function defined by the formula (78), $u(x)$, must be a solution for the Dirichlet problem (77).

Let us now prove that the homogeneous equation corresponding to (83),

$$\mu_0(s) + \int_S k(s,t)\mu_0(t)dt = 0 , \tag{84}$$

has no solution which are not zero identically,

In fact, let μ_0 be a solution of equation (84). The double layer potential $u_0(x)$ with density μ_0,

$$\mu_0(x) = -\frac{1}{2\pi}\int_S \frac{d}{d\nu_y}\log|y-x|\mu_0(y)ds_y , \tag{85}$$

satisfies the boundary condition

$$u_0^+(x^0) = 0 , \quad x^0 \in S ,$$

on account of (79) and (84); i.e., $u_0(x) = 0$ for all the points of the closed domain $D^+ \cup S$. In turn, it follows that

$$\left(\frac{\partial u^0}{\partial \nu}\right)^+ = 0 \tag{86}$$

on the whole of S.

By virtue of the property expressed by the equality (66) for a double layer potential, we find on the basis of (86) that

$$\left(\frac{\partial u_0}{\partial \nu}\right)^- = 0$$

on the whole of S. Since the normal derivative of the function $u_0(x)$, which is harmonic in D^- and is a double layer potential, equals to zero on $S = \partial D^-$ and $u_0(x) = 0$ on the whole of D^-. Hence $u_0^-(x^0) = 0$ on S.

Using the property expressed by the formulae (60) and (61) for a double layer potential, we obtain

$$\mu_0(x^0) = u_0^-(x^0) - u_0^+(x^0) = 0 \ .$$

Thus, the homogeneous equation (84) corresponding to (83) has no non-trivial solution. From this and on the basis of the well-known Fredholm's alternative theorem it follows that equation (83) for its arbitrary right-hand side, $\varphi \in \mathbf{C}^{0,0}(S)$, has one, and only one solution. By this the proof of the existence of the solution of the Dirichlet problem introduced in the previous statement of the problem follows.

The proof stated above shows the existence of solution of the Dirichlet problem (77) for a harmonic function was based on the possibility of reducing this problem to an equivalent Fredholm's integral equation of the second kind; to realize this possibility it is necessary to require something of the boundary S of the domain D^+ so that it is a closed Jordan curve with continuous curvature. If we reduce this requirement pertaining to the kernel $k(s,t)$ of the integral equation (83) may appear to have discontinuity which violates the Fredholm property of this equation. Nevertheless, this overstated smoothness of S is not necessary for existence of solution to the Dirichlet problem at all.

3.5.2. We can apply a method based on the theory of analytic functions of a complex variable to investigate this problem in a more general formulation: For a domain $D^+ \subset E_2$ with boundary S which is a closed Jordan curve, it is required to determine a harmonic function $u(x)$ which is continuous on $D^+ \cup S$ and satisfying the boundary condition (77).

We suppose that D^+ is a domain in the complex plane of variable $z = x_1 + ix_2$. For a harmonic function $u(x_1, x_2)$ we denote it by the notation $u(x_1, x_2) = u(z)$ and write the boundary condition (77) in the form

$$u(t) = \varphi(t) \ , \quad t \in S \ . \tag{87}$$

Let $F(z)$ be an analytic function in the domain D^+ and the real part of which is $u(z)$, i.e., Re $F(z) = u(z)$ for $z \in D^+$.

From the well-known conformal mapping principle (*Riemann's theorem*), we see that there is an *analytic function* $\zeta = f(z)$ which conformally maps the domain D^+ on the circle $d : |\zeta| < 1$ in the plane of the *complex variable* $\zeta = \xi + i\eta$: moreover, this function establishes a one-to-one continuous correspondence between the boundary S of the domain D^+ and the circumference $|\zeta| = 1$.

The function $F[f^{-1}(\zeta)]$ is analytic in the circle d, and its real part $u_*(\zeta) = \text{Re } F[f^{-1}(\zeta)]$ in d is a harmonic function satisfying the boundary condition

$$u_*(\tau) = h(\tau) , \quad |\tau| = 1 , \tag{88}$$

where $h(\tau) = \varphi[f^{-1}(\tau)]$.

The function $u_*(\zeta)$ which is harmonic in d and satisfying the boundary condition (88) is given by Poisson's formula

$$u_*(\zeta) = \frac{1}{2\pi} \int_{|\tau|=1} \frac{1 - |\zeta|^2}{|\tau - \zeta|^2} h(\tau) ds_\tau . \tag{89}$$

Therefore, there exists a solution for the Dirichlet problem formulated above, and it is expressed by means of $u_*(\zeta)$ in the form $u(z) = u_*[f(z)]$.

Since the harmonic function defined by the formula (89) has the given values (88) on the circumference ∂d from within the circle d, due to the one-to-one and continuous correspondence established by the function $\zeta = f(z)$ between $D \cup S$ and $d \cup \partial d$, we conclude that the function $u(z)$ has the given values (87) on the boundary S from within the domain D^+.

When $n > 2$ this property of the solution of the Dirichlet problem for harmonic functions will not be valid if the smoothness of the boundary S of the domain D^+ is broken (in particular, if the surface S has cusp points and cuspidal edges).

As already noticed in §2.7 of Chapter II, to prove the existence of a Green function $G(x, y)$ it is sufficient to solve the Dirichlet problem for the function $g(x, y)$ which is harmonic in the domain D^+ and satisfies the boundary condition of a special form

$$g(x, y) = -E(x, y) , \quad x \in S = \partial D^+ . \tag{90}$$

Starting from this it seems possible to expect that when $n = 2$ for a wide class of domains D (for example, their boundary are closed Jordan curves) the proof of the existence of a Green function can be obtained only by the indicated method. This is true in principle, but we are not always able to judge even the existence of the partial derivatives of G in $D^+ \cup S$ because of the special form of the boundary condition (90) (the elementary solution has singularity at $x = y$). When the boundary S of the domain D^+ is a closed Jordan curve with a continuous curvature, the Green function obviously exists, and its normal derivative is continuous on S.

3.5.3. When the Green function $G(x, y)$ of the Dirichlet problem for harmonic functions exists, we can repeat the reasoning made for reducing the formula (31) and hence verify that the formula

$$u(x) = -\frac{1}{\omega_n} \int_{D^+} G(x,t) f(t) d\tau_t \ , \tag{91}$$

where the function $f \in \mathbf{C}^{1,0}(D^+ \cup S)$, gives a particular solution $u(x)$ which is regular in the domain D^+ of Poisson's equation

$$\Delta u = f(x) \ , \quad x \in D^+ \tag{92}$$

Now we show that under the requirement that the domain D^+ is bounded the function $u(x)$ expressed by the formula (91) must satisfy the homogeneous boundary condition

$$\lim_{\substack{x \to x^0 \\ x \in D^+}} u(x) = 0 \ . \tag{93}$$

We cannot take the limit of the right-hand side of the formula (91) under the integral sign because we do not know whether or not the function $G(x, t)$ tends to zero uniformly with respect to $t \in D$ when $x \to x^0 \in S$. Therefore we can proceed as follows. We represent the function $u(x)$ in the form

$$u(x) = -\frac{1}{\omega_n} \int_{D_\varepsilon} G(x,t) f(t) d\tau_t - \frac{1}{\omega_n} \int_{d\varepsilon} G(x,t) f(t) d\tau_t \ ,$$

where $d_\varepsilon = D^+ \cap \{|x - x^0| < \varepsilon\}$ and D_ε is the part of D^+ that is outside the ball $|t - x^c| \leq \varepsilon$ with centre at the point x^0 and a sufficiently small radius ε.

Evidently,

$$\lim_{x \to x^0} \int_{D_\varepsilon} G(x,t) f(t) d\tau_t = \int_{D_\varepsilon} \lim_{x \to x_0} G(x,t) f(t) d\tau_t = 0 \ .$$

If we can show that for all $x \in d_\varepsilon$ the estimate

$$\int_{d_\varepsilon} G(x,t)d\tau_t < N(\varepsilon) ,$$

where $\lim_{\varepsilon \to 0} N(\varepsilon) = 0$ holds, then the validity of the equality (93) will be proved on account of the boundedness of the function $f(x)$.

By S_R we denote the sphere $|y-t| = R$ with centre at $t \in D^+$ and a sufficiently large radius R so that for any $t \in D^+$ thc sphere always encloses the domain D^+. Let

$$\Omega(x,t) = E(x,t) - E(y,t) , \quad |y-t| = R .$$

It is clear that the function $\Omega(x,t) > 0$ in the ball $|x-t| < R$, and

$$G(x,t) - \Omega(x,t) < 0$$

on the boundary S of the domain D^+. From this and taking into account the harmonicity of $G(x,t) - \Omega(x,t)$ on the domain D^+, we conclude that the $\Omega(x,t) > G(x,t) \geq 0$ in the whole of the domain D^+ by the extremum principle. Now the following interesting estimate

$$\int_{d_\varepsilon} G(x,t)d\tau_t < \int_{d_\varepsilon} \Omega(x,t)d\tau_t < N(\varepsilon) , \quad x \in d_\varepsilon$$

follows immediately from the inequality we have just obtained, since the integral over the domain d_ε of $\Omega(x,t)$ is uniformly and obsolutely convergent.

Note that if $u(x)$ is a solution of the homogeneous problem (93) for Poisson's equation (92) in the domain D^+ and $v(x)$ is a function harmonic in this domain and satisfies the nonhomogeneous boundary condition

$$v(x) = \varphi(x) , \quad x \in S ,$$

then the function $w(x) = u(x) + v(x)$ will be the solution of Poisson's equation (92), and satisfy the nonhomogeneous boundary condition $w(x) = \varphi(x), x \in S$.

3.5.4. In our discussion of the Dirichlet problem for harmonic functions we require the desired harmonic function $u(z) \equiv u(x_1, x_2)$ to be continuous on $D^+ \cup S$, hence, we also require the function $\varphi(x)$ given on S to be in the class

$\mathbf{C}^{0,0}(s)$. But in applications this requirement has occasionally to be weakened in alternative versions. Now we turn our attention to one of these.

For a domain $D^+ \subset E_2$ in the class $A^{1,h}$ the Dirichlet problem (87), stated as follows, is well-posed: to find a function $u(z)$ which is harmonic in the domain D^+, continuous on $D^+ \cup S$ except for some finite number of points $t_k \in S, k = 1, 2, \ldots, N$, where there may exist discontinuity of the first kind for the desired solution or it may tend to infinity in a way weaker than $\log|z - t_k|$ as $z \to t_k$ by assumption.

Uniqueness of solution for this problem is an immediate corollary of the following assertion: if the boundary values from within D^+ on S of a function $u(z)$ which is harmonic in the domain D^+ are equal to zero except for a finite number of points $\{t_k \in S\}, k = 1, 2, \ldots, N$, and

$$\lim_{\substack{z \to t_k \\ z \in D}} \frac{u(z)}{\log|z - t_k|} = 0 \ , \quad k = 1, 2, \ldots, N \ , \quad z = x_1 + ix_2 \ , \tag{94}$$

then $u(z) = 0$ in the whole of D^+.

To show the validity of this assertion we denote by D_δ the part of the domain D^+ that lies outside the closed circles with a sufficiently small radius δ, $|z - t_k| \leq \delta$, $k = 1, 2, \ldots, N$, and introduce for consideration a function which is positive and harmonic in D_δ,

$$v(z) = \varepsilon \sum_{k=1}^{N} \log \frac{d}{|z - t_k|} \ , \tag{95}$$

where d is the diameter of the domain D^+ and ε an arbitrary positive number.

Beginning with a definite value of $\delta > 0$, for any $\varepsilon > 0$ the boundary values on S_δ of D_δ of the functions $v(z) - u(z)$ and $v(z) + u(z)$ which are harmonic in this domain are positive. Therefore by using the extremum principle for harmonic functions we conclude that, for any point $z \in D_\delta$.

$$|u(z)| < v(z) \tag{96}$$

Since every fixed point $z \in D^+$ lies in a D_δ if it begins with a definite value $\delta > 0$, it follows from (96) that $u(z) = 0$ due to the arbitrariness of ε, so what we need has been proved.

For proving the existence of solution for the Dirichlet problem (87) stated above, we suppose that D^+ is a unit circle $|z| < 1$ in the plane of complex variable $z = x + iy$. This does not limit the generality because there exists an analytic function $\zeta = f(z)$ which conformally maps the domain D^+ on the unit circle $|\zeta| < 1$, belonging at least to the class $\mathbf{C}^{o,h}(|\zeta| \leq 1)$. Therefore the original problem has been reduced to determining the function $u_*(\zeta) = u[f_1^{-1}(\zeta)]$ which is harmonic in the circle $|\zeta| < 1$. The latter also has discontinuity in its boundary condition like the function $u(z)$.

The Poission's formula

$$u(z) = \frac{1}{2\pi} \int_{|t|=1} \frac{1 - |z|^2}{|t - z|^2} f(\vartheta) d\vartheta \;, \quad \vartheta = \arg\, t \;, \tag{97}$$

which gives the solution for the Dirichlet's problem $u(t) = f(\vartheta), 0 \leq \vartheta \leq 2\pi$, has been derived in Chapter II under the assumption that $u(z)$ is continuous in the closed circle $|z| \leq 1$.

We will confine ourselves to discuss the case when $f(\vartheta)$ is continuous on the circumference $|e^{i\vartheta}| = 1$ everywhere except for a point $z_0 = e^{i\vartheta}$, where it has discontinuity of the first kind,

$$\lim_{\vartheta \to \vartheta_0 - 0} f(\vartheta) = f^- \;, \quad \lim_{\vartheta \to \vartheta_0 + 0} f(\vartheta) = f^+ \;, \quad f^- \neq f^+ \;. \tag{98}$$

By $u_0(z)$ we denote the harmonic function

$$u_0(z) = \frac{1}{\pi}(f^- - f^+) \arg(z - z_0) \;, \tag{99}$$

where $\arg(z - z_0)$ should be understood as the principal branch of this function.

Let $f_0(\vartheta) = u_0(e^{i\vartheta}), 0 \leq \vartheta \leq 2\pi$. It is obvious that

$$\begin{aligned} \lim_{\vartheta \to \vartheta_0 - 0} f_0(\vartheta) &= (1 + \alpha)(f^- - f^+) \;, \\ \lim_{\vartheta \to \vartheta_0 + 0} f_0(\vartheta) &= \alpha(f^- - f^+) \;, \end{aligned} \tag{100}$$

where $\pi\alpha$ is the angle made by the positive direction of the tangent to the circumference $|e^{i\vartheta}| = 1$ at the point $e^{i\vartheta_0}$ with the positive direction of the real axis $y = 0$ in the plane of the complex variable $z = x + iy$.

The function $\varphi(\vartheta) = f(\vartheta) - f_0(\vartheta), 0 \leq \vartheta \leq 2\pi$, is continuous everywhere except for the point z_0, and at the point z_0 we have

$$\lim_{\vartheta \to \vartheta_0 - 0} \varphi(\vartheta) = \lim_{\vartheta \to \vartheta_0 + 0} \varphi(\vartheta) = (1 + \alpha) f^+ - \alpha f^- \;, \tag{101}$$

on account of (98) and (100).

Therefore after extending the definition of the function $\varphi(\vartheta)$ at z_0 by

$$\varphi(\vartheta) = \lim_{\vartheta \to \vartheta_0} \varphi(\vartheta) ,$$

it becomes continuous everywhere on the circumference $|t| = 1$.

In order to construct a function $u_1(z)$ which is harmonic in the circle $|z| < 1$, continuous in the closed circle $|z| \leq 1$, having the value $\varphi(\vartheta)$ on the circumference $|t| = 1$, we can utilize the Poisson formula (97),

$$u_1(z) = \frac{1}{2\pi} \int_{|t|=1} \frac{1 - |z|^2}{|t - z|^2} \varphi(\vartheta) d\vartheta . \tag{102}$$

The function $u(z)$ defined by the formula

$$u(z) = u_1(z) + u_0(z) , \tag{103}$$

which is harmonic in the circle $|z| < 1$, is continuous in the closed circle $|z| \leq 1$ everywhere except for the point z_0. Moreover,

$$\lim_{\substack{z \to t_0 \\ |z|<1}} u(z) = f(\varphi) , \quad t_0 \neq z_0 , \quad |t_0| = 1 .$$

When the point z tends to the point z_0 which is a discontinuous point of the function $f(\vartheta)$ from within the circle $|z| < 1$ and along the ray $\arg(z - z_0) = \pi\beta$, where $\pi\beta$ is the angle made by the vector $z - z_0$ and the positive direction of the real axis, we have from (99), (102) and (103)

$$\lim_{z \to z_0} u(z) = (1 - \lambda) f^+ + \lambda f^- ,$$

where $\lambda = \beta - \alpha$.

If the function $f(\vartheta)$ has a finite set of points of the first kind discontinuity, a similar situation will arise. As to the uniqueness of solution for the Dirichlet problem, in this case, with the boundary condition $\lim\limits_{z \to t} = f(\vartheta), t = e^{i\vartheta}, 0 \leq \vartheta \leq 2\pi$, it has been proved for the class of functions that are bounded on the closed circle $|z| \leq 1$.

It can be proved that if the function $f(\vartheta)$ given on the circumference $|t| = 1$ is integrable, then the formula (97) gives a harmonic function $u(z)$ on the circle $|z| < 1$ which tends to $f(\vartheta)$ as $|z| \to 1$ for almost all $\vartheta, 0 \leq \vartheta \leq 2\pi$.

Notice that if $t'' = ei\vartheta'', t' = e^{i\vartheta'}, \vartheta'' < \vartheta'$ are points on $|t| = 1$, then the expression

$$u(\vartheta, \vartheta''; z) = \frac{1}{\pi} \arg \left[\frac{z - t'}{z - t''} e^{i\frac{\vartheta'' - \vartheta'}{2}} \right] ,$$

selecting only its principal branch, is a harmonic function in the circle $|z| < 1$. It is equal to one on the arc $\vartheta'' < \vartheta < \vartheta'$ of the circumference $|t| = 1$ and equal to zero on the complement of the closed arc $\vartheta'' \leq \vartheta \leq \vartheta'$ with respect to the whole circumference $|t| = 1$. It is said to be the *harmonic measure* of the arc $t''t'$ of the circumference at the point $z, z < 1$, with respect to the unit circle $|z| < 1$.

§3.6 Neumann's Problem. Problem with a Directional Derivative

3.6.1. Let D be a bounded domain in the space E_n, belonging to the class $A^{1,h}$, and $f, l_1, \ldots, l_n$ be real continuous functions given on its boundary $S = \partial D$ with $\sum_{i=1}^{n} l_i^2 \neq 0$ everywhere on S.

By *Neumann's problem* for harmonic functions we mean a problem as follows: to find a function $u(x)$ which is harmonic in the domain D, being of the class $\mathbf{C}^{1,0}(D \cup S)$, and satisfying the boundary condition

$$\frac{\partial u}{\partial \nu_x} = f(x) , \quad x \in S , \tag{104}$$

This problem is a special case of the so-call directional derivative problem: to find a function $u(x)$ which is harmonic in the domain D being of the class $\mathbf{C}^{1,0}(D \cup S)$, and satisfies the boundary condition

$$l(x) \cdot \text{grad}\, u(x) = f(x) , \quad x \in S , \tag{105}$$

where the vector $l = (l_1, \ldots, l_n)$.

Without loss of generality we can suppose that $|l| = 1$ because $l \neq 0$.

As it has already been proved in §7 of Chapter II, for any function $u(x)$ which is harmonic in the domain D, being in the class $\mathbf{C}^{1,h}(D \cup S)$, the equality

$$\int_S \frac{\partial u(x)}{\partial \nu_x} ds_x = 0 ,$$

holds, and if $\frac{\partial u(x)}{\partial \nu_x} = 0$ everywhere on S, then $u(x) = \text{Const.}$ in $D \cup S$.

On the basis of these properties of harmonic functions we conclude that if the Neumann problem (104) has a solution $u(x)$, then (1) it is determined up to an arbitrary constant term and (2) the integral over the boundary of the domain D of $f(x)$ equals to zero,

$$\int_S f(x)ds_x = 0 . \tag{106}$$

Therefore, the equality (106) is a necessary condition for solvability of the Neumann problem (104).

3.6.2. If we seek a solution of the Neumann problem in the form of simple layer potential (55),

$$u(x) = \frac{1}{\omega_n}\int_S E(x,y)\mu(y)ds_y ,$$

then

$$\left[\frac{\partial u(x)}{\partial \nu_x}\right]^+ = \frac{1}{2}\mu(x) + \frac{\partial u(x)}{\partial \nu_x} , \quad x \in S ,$$

by virtue of its property (71), and to determine the unknown distribution density μ of mass we obtain the Fredholm's integral equation of the second kind,

$$\mu(x) + \int_S k_*(x,y)\mu(y)ds_y = 2f(x) , \quad x \in S , \tag{107}$$

whose kernel is given by the formula

$$k_*(x,y) = \frac{2}{\omega_n}\frac{\partial}{\nu_x}E(x,y) .$$

As we have already proved, when $n = 2$ and the boundary S of the domain D is a closed Jordan curve with a continuous curvature, the function $k_*(x,y)$ must be continuous in all of the variables $x, y \in S$. It can be proved that when $n > 2$ and if the domain D is in the class $A^{1,h}$, the Fredholm theory is also applicable to the integral equation (107). In contrast to the Dirichlet problem in the case of the Neumann problem the corresponding homogeneous integral equation of (107) has nontrivial solutions, but the fulfilment of the condition

(106) guarantees the solvability of the integral equation (107), and hence that of Neumann's problem. Thus, (106) is a sufficient and necessary condition for the existence of the solution for this problem.

3.6.3. The solution of Neumann's problem (104) for harmonic functions can be represented as a quadrature for the case of a circle if the condition (106) is fulfilled.

We can actually confine ourselves to observing a unit circle $D : |z| < 1, z = x_1 + ix_2$. Since the equalities $x_1 = \cos \hat{\nu x_1}$ and $x_2 = \cos \hat{\nu x_2}$ hold on the circumference $|z| = 1$, the boundary condition (104) can be written in the form

$$x_1 \frac{\partial u}{\partial x_1} + x_2 \frac{\partial u}{\partial x_2} = f(z) , \quad |z| = 1 ,$$

or

$$\operatorname{Re} [z\phi'(z)] = f(z) , \quad |z| = 1 , \tag{108}$$

where $\phi'(z) = \frac{\partial u}{\partial x_1} - i \frac{\partial u}{\partial x_2}$ is the derivative of the function $\phi(z) = u(z) + iv(z)$ which is analytic in D.

The function $z\phi'(z)$, which is analytic in the circle $|z| < 1$ and the real part of whose boundary values on the circumference $|z| = 1$ is given by (108), is given by the Schwarz's formula (107) in Chapter II,

$$z\phi'(z) = \frac{1}{\pi_i} \int_{|t|=1} \left(\frac{1}{t - z} - \frac{1}{2t} \right) f(t) dt + iC_0 , \tag{109}$$

where C_0 is an arbitrary real constant.

Supposing that the condition (106) is fulfilled, i.e.,

$$\frac{1}{i} \int_{|t|=1} \frac{f(t)}{t} dt = \int_{|t|=1} f(\vartheta) d\vartheta = 0 , \quad e^{i\vartheta} = t ,$$

we can give the equality (109) the form

$$z\phi'(z) = \frac{z}{\pi} \int_{|t|=1} \frac{f(t)}{t - z} d\vartheta + iC_0 .$$

Thus, $C_0 = 0$ and hence

$$\phi(z) = -\frac{1}{\pi} \int_{|t|=1} f(t) \log \left(1 - \frac{z}{t} \right) d\vartheta + C_1 , \tag{110}$$

where C_1 is an arbitrary constant and $\log(1-\frac{z}{t})$ should be considered as the branch of this function that is equal to zero at $z=0$.

Rewriting formula (110) in the form

$$\phi(z)=-\frac{1}{\pi}\int_{|t|=1}\log\ |t-z|f(t)d\vartheta+\frac{i}{\pi}\int_{|t|=1}\arg\ (1-\frac{z}{t})f(t)d\vartheta+c_1$$

and singling the real part out for the desired solution $u(z)$ of Neumann's problem, we obtain

$$u(z)=-\frac{1}{\pi}\int_{|t|=1}\log\ |t-z|f(\vartheta)d\vartheta+C\ , \tag{111}$$

where C is an arbitrary real constant.

3.6.4. When $n>2$, it is also possible to write out the solution of Neumann's problem (104) for harmonic functions in a ball of the Euclidean space E_n in the form of a quadrature.

In fact, in the case of a ball $D:|x|<1$ the boundary condition (104) has the form

$$x\ \text{grad}\ u(x)=f(x)\ ,\quad x\in S\ ,\quad S:|x|=1\ . \tag{108_1}$$

Along with the $u(x)$ the function

$$x\ \text{grad}\ u(x)=v(x)\ ,\quad x\in D\ , \tag{109_1}$$

is also harmonic in D, for which (108_1) is the condition for Dirichlet's problem $v(x)=f(x), x\in S$. We can obtain the function $v(x)$ by using Poisson's formula (96) in Chapter II.

$$v(x)=\frac{1}{\omega_n}\int_{|y|=1}\frac{1-|x|^2}{|y-x|^n}f(y)ds_y\ ,\quad x\in D\ .$$

From this and taking into consideration that when $x=0$ we should have $v(0)=0$ due to (109_1), we again arrive at the condition (106) which is necessary for the existence of solution for Neumann's problem. If this condition is satisfied then the desired solution of the problem (104) is given by the formula

$$u(x)=\int_0^1 v(tx)\frac{dt}{t}+C\ , \tag{110_1}$$

where C is an arbitrary constant.

If fact, because $v(0) = 0$ we obtain from (110_1)

$$U_{x_i} = \int_0^1 v_{\xi_i} dt \ , \quad u_{x_i x_i} = \int_0^1 v_{\xi_i \xi_i} t dt \ ,$$
$$x \text{ grad } u(x) = v(x) \ , \quad \Delta u = \int_0^1 \delta v t dt \ ,$$

as a result of the replacement of variables, $\xi = tx$, and differentiation, and hence the correctness of the stated assertion follows.

Now write the equality (110_1) in the form

$$u(x) = \frac{1}{\omega_n} \int_S N(x,y) f(y) ds_y + C \ , \tag{111_1}$$

where

$$N(x,y) = \int_0^1 Q_n(t) dt$$

and

$$Q_n(t) = \frac{1}{tR^{\frac{n}{2}}} - \frac{tx^2}{R^{\frac{n}{2}}} - \frac{1}{t} \ , \quad Q(0) = \lim_{t \to 0} Q(t) \ ,$$
$$R(t) = 1 - 2xyt + x^2 t^2 \ , \quad |y| = 1 \ , \quad xy = \sum_{i=1}^n x_i y_i \ .$$

By simple computations we find

$$N(x,y) = \begin{cases} -2 \log |y - x| \ , & n = 2 \\ \dfrac{2}{|y-x|} - \log \dfrac{2}{1 - xy + |y - x|} \ , & n = 3 \\ \dfrac{1}{|y-x|^2} - \log |y - x| + \dfrac{xy}{\delta^{\frac{1}{2}}} \operatorname{arctg} \dfrac{\delta^{\frac{1}{2}}}{1 - xy} \ , & n = 4 \end{cases}$$

where $\delta = x^2 - (xy)^2$.

3.6.5. Let D be the half-plane $x_n > 0$ in the space E_n. The condition (104) of Neumann's problem this time has the form

$$\frac{\partial u}{\partial x_n} = -f(x) \ , \quad x = (x_1, \dots, x_{n-1}, 0) \ . \tag{108_2}$$

Since (108_2) is the condition of the Dirichlet problem for the function u_{x_n} which is harmonic in D, we have

$$u_{x_n} = -\frac{\Gamma(\frac{n}{2})}{\pi^{\frac{1}{2}}} x_n \cdot \int_{y_n=0} \frac{f(y)dy\ldots dy_{n-1}}{[\sum(y_i - x_i)^2 + x_n^2]^{\frac{n}{2}}}$$

according to the formula (104) in Chapter II if it is bounded, and hence for the desired solution of the problem (104) we obtain

$$u(x) = \frac{\Gamma(\frac{n}{2})}{(n-2)\pi^{\frac{n}{2}}} \int_{y_n=0} \frac{f(y)dy\ldots dy_{n-1}}{[\sum(y_i - x_i)^2 + x_n^2]^{\frac{n-2}{2}}} .$$

3.6.6. We now turn to discuss the problem with a directional derivative (105). The theory for this problem has been detailed only for two-dimensional domains. Here we shall confine ourselves to discussing this case.

Let D be a bounded domain in the plane of the complex variable $x = x_1 + ix_2$ for whose boundary a closed Lyapnov curve S is taken. The problem with a directional derivative, which we will discuss below, is stated as follows: determine a function $u(z) \equiv u(x_1, x_2)$ which is harmonic in the domain D and of the class $\mathbf{C}^{1,h}(D \cup S)$ by using the boundary condition

$$l_1(t)\frac{\partial u}{\partial x_1} + l_2(t)\frac{\partial u}{\partial x_2} = f(t) , \quad x_1 + ix_2 = t \in S , \tag{112}$$

where l_1, l_2 and f are given real functions of the class $\mathbf{C}^{0,h}(S)$, and $l_1^2 + l_2^2 = 1$.

We denote by $\phi(z)$ the function that is analytic in the domain D and its real part is a desired solution $u(z)$ for the problem (112). Because $\phi'(z) = \frac{\partial u}{\partial x_1} - i\frac{\partial u}{\partial x_2}$, the condition (112) can be written in the form

$$\text{Re}\,[\varphi(t)F(t)] = f(t) , \quad t \in S , \tag{113}$$

where

$$\varphi(t) = l_1(t) + il_2(t) , \tag{114}$$

$$F(z) = \phi'(z) . \tag{115}$$

Thus, the problem (112) has been reduced to the problem of determining the function $F(z)$ which is analytic in the domain D and of the class $\mathbf{C}^{0,h}$

$(D \cup S)$ by using the boundary condition (113). To discuss this problem, named *Hilbert's problem*, without loss of generality, we supposed that D is the unit circle $|z| < 1$. This can always be achieved by using a conformal mapping.

By n we denote the integer

$$n = \frac{1}{2\pi} Var\ [\arg \varphi(t)]_S \ , \tag{116}$$

where Var denotes, through the square brackets, the increment of the argument of the function $\varphi(t)$ when the point t on the circumference $S : |t| = 1$ moves round on it in a single-valued way and along the positive direction.

It is obvious that the function

$$\alpha(t) = \arg\ \varphi(t) - n\vartheta \ , \quad t = e^{i\vartheta} \ , \tag{117}$$

is single-valued and continuous in Hölder's sense on S.

A function $\psi(z)$ which is analytic in the circle D, being of $\mathbf{C}^{0,h}(D \cup S)$, and satisfies the boundary condition

$$\text{Re}\ \psi(t) = \alpha(t) \ , \tag{118}$$

can be constructed by Schwarz's formula as

$$\psi(z) = \frac{1}{\pi i} \int_S \frac{\alpha(t)dt}{t - z} - \frac{1}{2\pi i} \int_S \frac{\alpha(t)dt}{t} \ ,$$

provided that

$$\text{Im}\psi(0) = 0 \ . \tag{119}$$

Once the function $\psi(z)$ is found, the boundary condition (113) can be rewritten in the form

$$\text{Re}\ [t^n e^{i\psi(t)} F(t)] = e^{-\text{Im}\psi(t)} f(t) \ , \quad t \in S \ , \tag{120}$$

on the basis of (114), (117) and (118).

First of all, we observe the case of $n > 0$. The function

$$\Omega(z) = z^n e^{i\psi(z)} F(z) \ , \tag{121}$$

which is analytic in the circle D and of the class $\mathbf{C}^{0,h}(D \cup S)$, and satisfies the boundary condition (120), is given by Schwarz's formula

$$\Omega(z) = \frac{1}{\pi i} \int_S e^{-\mathrm{Im}\psi(t)} \left(\frac{1}{t-z} - \frac{1}{2\pi} \right) f(t) dt + iC \ . \tag{122}$$

From (121) and (122) we obtain

$$F(z) = z^{-n} e^{i\psi(z)} \left[\frac{1}{\pi i} \int_S e^{-\mathrm{Im}\ \psi(t)} \left(\frac{1}{t-z} - \frac{1}{2t} \right) f(t) dt + iC \right] \ . \tag{123}$$

Since the analytic function $e^{-i\psi(z)}$ nowhere vanishes on $D \cup S$, the function $F(z)$ expressed by the formula (123) will be analytic in the domain D as long as the analytic function in the brackets of the right-hand side of this formula has a zero of not less than n-th order at $z = 0$, i.e., when $c = 0$

$$\int_S e^{-\mathrm{Im}\psi(t)} t^{-k-1} f(t) dt = 0 \ , \quad k = 0 \ , \dots \ , n-1 \ . \tag{124}$$

Therefore, the problem (113) for the considered case has one and only one solution if the right-hand part $f(t)$ in the boundary condition (113) satisfies the additional requirements (124). They represent $2n-1$ actual integral conditions which the function f needs to satisfy for solvability of the problem (113).

Now suppose $n \leq 0$. We represent the desired function $F(z)$ satisfying the condition (120) in the form

$$F(z) = \sum_{k=0}^{-n-1} \frac{1}{k!} F^{(k)}(0) z^k + z^{-n} F_1(z) \ , \tag{125}$$

where $F_1(z)$ is also an unknown function and it is supposed that $F_1(z) = F(z)$ when $n = 0$.

Taking into account (125), the boundary condition (120) can be written in the form

$$\mathrm{Re}[e^{i\psi(t)} F_1(t)] = e^{-\mathrm{Im}\psi(t)} f(t) - \mathrm{Re} \left[e^{i\psi(t)} \sum_{k=0}^{-n-1} \frac{1}{k!} F^{(k)}(0) t^{k+n} \right] \ . \tag{126}$$

Applying Schwarz's formula again, we have

$$e^{i\psi(z)} F_1(z) = iC + \frac{1}{\pi i} \int_S \left\{ e^{-\mathrm{Im}\psi(t)} f(t) - \mathrm{Re} \left[e^{i\psi(t)} \sum_{k=0}^{-n-1} \frac{t}{k!} F^{(k)}(0) \right] \right\}$$
$$\left(\frac{1}{t-z} - \frac{1}{2t} \right) dt \tag{127}$$

due to (126). Thus, we find from the formulae (125) and (127) a family of solutions $F(z)$ for the problem (113),

$$F(z) = \sum_{k=0}^{-n-1} \frac{1}{k!} E^{(k)}(0) z^k + iCe^{-i\psi(z)} z^{-n}$$
$$+ z^{-n} e^{-i\psi(z)} \frac{1}{\pi i} \int_S \left\{ e^{-\mathrm{Im}\psi(t)} f(t) - \mathrm{Re}\left[e^{i\psi(t)} \sum_{k=0}^{-n-1} \frac{t^{n+k}}{k!} F^{(k)}_{(0)} \right] \right\}$$
$$\left(\frac{1}{t-z} - \frac{1}{2t} \right) dt \ , \tag{128}$$

which contains a real number C and $-n$ of complex arbitrary constants $F^{(k)}(0)$, $k = 0, 1, \dots, -n-1$.

Substituting the value of $F(z)$ obtained from the formula (128) into (115) we obtain the function $\phi(z)$ by integration, and, hence, the function $u = \mathrm{Re}\ \phi(z)$ which is the solution of the problem (112). This expression obtained for $u(z)$ depends linearly on $-2n+2$ arbitrary real constants.

§3.7 Some Other Problems for Elliptic Equations

3.7.1. The boundary problems for elliptic equations have been posed not only for bounded domains.

Let S be a closed $(n-1)$-dimensional Lyapnov surface in space E_n, and D^- be a infinite domain with boundary S. The problem we will discuss below is stated as follows: Determine a regular function u which is harmonic in the domain D^-, being of the class $\mathbf{C}^{0,0}(D \cup S)$, and satisfies the boundary condition

$$u(y) = \varphi(y) \ , \quad y \in S \ , \tag{129}$$

where φ is a real continuous function given on S.

This problem, different from the Dirichlet problem for a finite domain $D^+ = C(D^- \cup S)$ (i.e. the interior problem), is naturally called the *exterior Dirichlet problem.*

According to definition the regularity of the function $u(x)$ which is harmonic in D^- means that when $|x| \to \infty$ this function tends to zero not slower than $|x|^{2-n}$ if $n > 2$, and tends to some finite limit if $n = 2$.

We suppose, without loss of generality, that the point $x = 0$ belongs to the domain D^+.

By taking the inversion

$$x' = \frac{x}{|x|^2} , \tag{130}$$

the domain D^- with boundary S turns to a domain D'^+ with boundary S' in the space E'_n of variables $x'_1, \ldots , x'_n$.

The function $v(x')$ defined by the formula

$$v(x') = |x'|^{2-n} u\left(\frac{x'}{|x'|^2}\right) \tag{131}$$

with supplementary definition, $\lim_{x' \to 0} v(x') = v(0)$, at the point $x = 0$ is harmonic in the domain D', and it belongs to the class $\mathbf{C}^{0,0}(D'^+ \cup S')$, satisfying the boundary condition

$$v(y') = |y'|^{2-n} \varphi\left(\frac{y'}{|y'|^2}\right) , \quad y' \in S' , \tag{132}$$

on account of (129).

Thus, the exterior Dirichlet problem for harmonic functions (129) has been reduced to an interior Dirichlet's problem discussed before, also for harmonic functions.

Taking into consideration that the equality (130) turns to

$$x = \frac{x'}{|x'|^2}$$

one-to-one if $x \neq 0$, we conclude that we can obtain the desired solution for the exterior Dirichlet's problem (129),

$$u(x) = |x|^{2-n} v\left(\frac{x}{|x|^2}\right) , \tag{133}$$

from the formula (131) as long as there exists a solution $v(x')$ for the Dirichlet problem (132).

3.7.2. When D^- is the exterior of the unit ball $|x| \leq 1$, the problem (129) can be reduced to the Dirichlet problem (132) for the unit sphere $|x'| < 1$ in the space E'_n by the replacement (130), and, furthermore,

$$v(y') = \varphi(y') , \quad |y'| = 1 .$$

Therefore, due to Poisson's formula [cf. the formula (96), Chapter II] we have

$$v(x') = \frac{1}{\omega_n} \int_{|y'|=1} \frac{1 - |x'|^2}{|y' - x'|^n} \varphi(y') ds_{y'} . \tag{134}$$

Since for $|y'| = 1$ we have the equality $|x||y' - x'| = |y - x|$ by virtue of the formula (130), for the desired solution $u(x)$ of the exterior problem (129) we obtain the formula (cf. (102), Chapter II)

$$u(x) = \frac{1}{\omega_n} \int_{|y|=1} \frac{|x|^2 - 1}{|y - x|^n} \varphi(y) ds_y .$$

The exterior Dirichlet problem (129) can be investigated without having to reduce it to an interior Dirichlet's problem. For example, if we are going to find a solution $u(x)$ of this problem in the form of a double layer potential (54), then in order to determine the unknown density μ, we obtain a Fredholm's integral equation of the second kind by the formula $u^-(x^0) - u(x^0) = \frac{1}{2}\mu(x^0)$, for which the univalent solvability can be established by repeating the argument stated before when we were discussing the interior Dirichlet problem.

3.7.3. An *exterior Neumann problem* is to determine a function $u(x)$ which is regularly harmonic in the domain D^- and of the class $\mathbf{C}^{1,h}(D^- \cup S)$ by using the boundary condition

$$\left(\frac{\partial u}{\partial \nu_x}\right)^- = \varphi(x) , \quad x \in S ,$$

where ν is the normal to S and φ a given real function of the class $\mathbf{C}^{0,h}(S)$.

One cannot simply reduce it to a similar problem with a bounded domain, as we have done for the exterior Dirichlet's problem. But if we seek its solution $u(x)$ in the form of a simple layer potential and make use of the formula $\left(\frac{\partial u}{\partial \nu_x}\right)^- - \frac{\partial u(x)}{\partial \nu_x} = -\frac{1}{2}\mu(x)$, then in order to determine the density μ we have a Fredholm's integral equation of the second kind which we can investigate to the end by applying the standard techniques introduced before.

3.7.4. In the case that the linear partial differential operator Lu in its specification domain D satisfies the condition for uniform ellipticity, a broad kind of boundary problems for equation

$$Lu = f(x) , \quad x \in D \tag{135}$$

can be covered by the following *Poincare's problem*: to find a solution $u(x)$ of equation (135) which is regular in the domain D, belonging to the class $\mathbf{C}^{1,h}(D \cup S)$, and satisfies the boundary condition

$$l(x) \cdot \text{ grad } u(x) + q(x)u(x) = r(x) , \quad x \in S , \tag{136}$$

where the real vector $l = (l_1, \ldots, l_n)$ and the real scalars q and r are given on the whole of S. If $l(x) = 0$ everywhere on S and q is nowhere zero, then the boundary condition

$$u(x) = g(x) , \quad g = \frac{r(x)}{q(x)} , \tag{137}$$

is obtained.

The problem (135)–(137) is usually called a *first boundary problem* or a *Dirichlet's problem*, as in the case of harmonic functions.

When $q(x) \equiv 0, x \in S$, the condition (135) has the form

$$l(x) \cdot \text{grad } u(x) = r(x) , \quad x \in S . \tag{138}$$

When the vector l is nowhere zero, the problem with the boundary condition (138) for the equation (135) is called a *problem with a directional derivative.*

If the vector in the boundary condition (138) coincides with the conormal N, then the problem with a directional derivative is called a *Neumann's problem* for the equation (135). Notice that if equation (135) is a Laplace's equation, then the conormal $\mathbf{N}$ coincides with the normal ν to the boundary S of the domain D.

Under the assumption that the corresponding homogeneous equation of (135),

$$Lu = 0 , \tag{139}$$

has a principal elementary solution $\Omega(x, y)$ for the regular solutions of the equation (139) we can obtain the formula

$$u(x) = \int_S a(y)\Omega(x, y)\frac{\partial u(y)}{\partial \mathbf{N}_y} - \int_S Q_y \Omega(x, y)u(y)ds_y ,$$

where $\mathbf{N}_y$ is the conormal,

$$a = \left\{ \sum_{i=i}^{n} \left(\sum_{j=1}^{n} A_{ij} \cos \hat{\nu y_j} \right)^2 \right\}^{\frac{1}{2}} ,$$

$$Q_y = a(y) \frac{\partial}{\partial \mathbf{N}_y} - b(y) , \quad b(y) = \sum_{i=1}^{n} e_i(y) \cos \hat{\nu y_i} ,$$

if we repeat the reasoning applied to the reduction of the integral representation of harmonic functions (87) in Chapter II. The functions

$$u(x) = \int_S \frac{d}{d\mathbf{N}_y} \Omega(x, y) \mu(y) ds_y$$

and

$$v(x) = \int \Omega(x, y) \mu(y) ds_y ,$$

which are solutions of equation (139) are called *generalized double layer* and *simple layer potentials*, respectively. This property makes it possible for us to reduce the Dirichlet problem and the Neumann problem of equation (139) to equivalent Fredholm's integral equations of the second kind, and to investigate these problems in the same way as we have done in the case of harmonic functions.

§3.8 Extremum Principle for Equation (139) and the Uniqueness of Solution for Problem (135)–(137)

3.8.1. In the theory of second order linear elliptic equation the extremum principle plays an important role: If

$$c(x) < 0 \tag{140}$$

everywhere in the domain D, then a solution $u(x)$ which is regular in this domain of the elliptic equation (139) cannot take, at a point $x \in D$, either a negative relative minimum or a positive relative maximum.

Supposing that the function $u(x)$ reached its negative minimum at some point $x \in D$, then at this point we would have

$$\frac{\partial u}{\partial x_i} = 0 \,, \quad i = 1, \dots, n \tag{141}$$

$$\sum_{i,j=1}^{n} \frac{\partial^2 u}{\partial x_i \partial x_j} \lambda_i \lambda_j \geq 0 \,, \tag{142}$$

where $\lambda_1, \dots, \lambda_n$ are arbitrary real parameters. Since we can always represent at the point $x \in D$ the positively definite form $\sum_{i,j=1}^{n} A_{ij} \lambda_i \lambda_j$ in the form

$$\sum_{i,j=1}^{n} A_{ij} \lambda_i \lambda_j = \sum_{k=1}^{n} \left(\sum_{l=1}^{n} g_{kl} \lambda_l \right)^2 \,,$$

for the coefficients A_{ij} the expression

$$A_{ij} = \sum_{s=1}^{n} g_{si} g_{sj} \,, \quad i, j = 1, \dots, n \tag{143}$$

holds.

On the basis of (142) and (143) we have

$$\sum_{i,j=1}^{n} A_{ij} \frac{\partial^2 u}{\partial x_i \partial x_j} = \sum_{i,j=1}^{n} \sum_{s=1}^{n} \frac{\partial^2 u}{\partial x_i \partial x_j} g_{si} g_{sj} \geq 0 \,. \tag{144}$$

Taking into account $u(x) < 0$ we obtain $Lu > 0$ because of (9), (140), (141) and (144); but this contradicts the equality (139). This contradiction refutes our assumption.

Similarly, it can be proved that the function $u(x)$ cannot arrive at its positive relative maximum.

3.8.2. It follows from the extremum principle that Dirichlet's problem (135)–(137) in a bounded domain D cannot have more than one solution if we require that $u(x) \in \mathbf{C}^{0,0}(D \cup S)$. In fact, for the difference $u = u_1 - u_2$ of any two solutions u_1 and u_2 of this problem we must have

$$Lu(x) = 0 \,, \quad x \in D \,, \quad u(y) = 0 \,, \quad y \in S \,. \tag{145}$$

Because of $\max\limits_{x\in S}|u(x)| = 0$, it follows from the extremum principle that in the whole domain $D, u(x) = 0$, i.e. $u_1(x) \equiv u_2(x)$.

3.8.3. It is easy to see that when the upper bound of $a(x,y)$ in a bounded domain D is finite the uniqueness of solution for the problem (137) of the equation

$$Lu \equiv \Delta u + au_x + bu_y + cu = f \tag{146}$$

holds provided that the condition

$$c(x,y) \leq 0 \tag{147}$$

is also satisfied.

To verify this, it is sufficient to show that under the condition (147) the homogeneous problem (145) has no solution which is different from the identical zero.

As a result of the replacement of the function $u(x,y)$ defined by the formula $u = (A - e^{-\mu x})v(x,y)$, where $A =$ constant > 0 and the positive number μ is larger than the upper bound of $a(x,y)$ in the domain D, we obtain by (145) for v,

$$\begin{aligned} \Delta v + a_1 v_x + bv_y + c_1 v = 0 \ , \quad (x,y) \in D \ , \\ v(x,y) = 0 \ , \quad (x,y) \in S \ , \end{aligned} \tag{148}$$

where

$$a_1 = a + 2\mu(Ae^{\mu x} - 1)^{-1} \ , \quad c_1 = c - \mu(\mu - a)(Ae^{\mu x} - 1)^{-1} \ .$$

We can choose the number A such that the expression $Ae^{\mu x} - 1$ is positive and hence the condition $C_1(x,y) < 0$ is satisfied in the whole of D. For this case, as we have already proved, the problem (148) has no solution which is different from the identical zero, i.e., the solution $u(x,y)$ of the homogeneous problem (145) is identically equal to zero in the domain D.

Chapter IV. Generalized Solutions of Partial Differential Equations

§4.1 A Brief Survey of Methods for Constructing Classical Solutions

4.1.1. The solutions of partial differential equations which satisfy some boundary, initial, mixed, etc., conditions corresponding to the discussed problem in the usual sense are said to be *classical solutions*, i.e., at every point of the supports, sufficient smoothness has been assumed for them.

Among the methods of constructing classical solutions of partial differential equations, the *method of separation of variables* is universal and the first historically.

The essence of this method can be observed through the example of linear partial differential second order equation of the form

$$\sum_{i,j=1}^{n} A_{ij}(x)\frac{\partial^2 u}{\partial x_i \partial x_j} + \sum_{i=1}^{n} B_i(x)\frac{\partial u}{\partial x_i} + C(x)u$$
$$= \alpha(t)\frac{\partial^2 u}{\partial t^2} + \beta(t)\frac{\partial u}{\partial t} + \gamma(t)\mu \ , \tag{1}$$

where the coefficients A_{ij}, B_i and C are fixed functions of spatial variables $x_1, x_2, \dots, x_n$, and α, β, γ are functions of variable t which usually plays the role of time.

If we seek a solution $u(x,t)$ of equation (1) as a product of two functions $v(x)$ and $w(t)$,

$$u(x,t) = v(x)w(t) \ , \tag{2}$$

then

$$w(t)\left[\sum_{i,j=1}^{n} A_{ij}(x)\frac{\partial^2 v}{\partial x_i \partial x_j} + \sum_{i=1}^{n} B_i(x)\frac{\partial v}{\partial x_i} + C(x)v\right]$$
$$= v(x)\left[\alpha(t)\frac{d^2 w}{dt^2} + \beta(t)\frac{dw}{dt} + \gamma(t)w\right] .$$

The necessary and sufficient conditions to make the equality true for $x = (x_1, \dots, x_n)$ and t in the domain of the given equation (1) are to satisfy two equalities

$$\sum_{i,j=1}^{n} A_{ij}(x)\frac{\partial^2 v}{\partial x_i \partial x_j} + \sum_{i=1}^{n} B_i(x)\frac{\partial v}{\partial x_i} + [c(x) + \lambda]v = 0 \tag{3}$$

and

$$\alpha(t)\frac{d^2 w}{dt^2} + \beta(t)\frac{dw}{dt} + [\alpha(t) + \lambda]w = 0 \ , \tag{4}$$

where λ is a constant, and

$$\begin{aligned} \lambda &= -\frac{1}{v(x)}\left[\sum_{i,j=1}^{n} A_{ij}(x)\frac{\partial^2 v}{\partial x_i \partial x_j} + \sum_{i=1}^{n} B_i(x)\frac{\partial v}{\partial x_i} + C(x)v\right] \\ &= -\frac{1}{w(t)}\left[\alpha(t)\frac{d^2 w}{\partial t^2} + \beta(t)\frac{dw}{dt} + \gamma(t)w\right] \ . \end{aligned}$$

Thus, the independent variables x, t and the functions $v(x)$, $w(t)$ have been separated, and to determine them we obtain the partial differential equation (3) (if $n > 1$) where the number of independent variables is less than that of equation (1) by one, and the ordinary differential equation (4).

In many cases of the equation (1) the choice of its solutions in the form (2) makes it possible to investigate the considered problem in exhaustive detail.

4.1.2. Here we only discuss the *oscillation problems of an elastic membrane* with the boundary fixed along the curve S situated in the plane of the variables x and y. Suppose that this membrane is present in a finite domain G when it is in the equilibrium position and $S = \partial G$. It is well-known that the discussed process, as described by a function $u(x, y, t)$, is the regular solution of the wave equation with two spatial variables x and y

$$\frac{\partial^2 u}{\partial t^2} - \frac{\partial^2 u}{\partial x^2} - \frac{\partial^2 u}{\partial y^2} = 0 \ . \tag{5}$$

By the *first basic mixed problem* for the equation (5) we mean the problem to determine the regular solution $u(x, y, t)$ of this equation, which satisfies the initial conditions

$$u(x, y, 0) = \varphi(x, y) \ , \quad \frac{\partial u(x, y, 0)}{\partial t} = \psi(x, y) \ , \quad (x, y) \in G \ , \tag{6}$$

and the boundary condition

$$u(x,y,t) = 0 \ , \quad (x,y) \in S \ , \quad t \geq 0 \ . \tag{7}$$

The semicylinder $Q = \{G \times (0 < t < \infty)\}$ is the domain of the definition of the solution for the problem (5)–(6)–(7).

4.1.3. The method of separation of variables has been used to investigate this problem. In the considered case instead of the equations (3) and (4) the *Helmholtz equation*

$$\Delta v(x,y) + \lambda v(x,y) = 0 \ , \quad \frac{\partial^2}{\partial x^2} + \frac{\partial^2}{\partial y^2} = \Delta \ , \quad (x,y) \in G \lambda = \text{const.} \ , \tag{8}$$

and the ordinary differential equation

$$W''(t) + \lambda W(t) = 0 \ , \quad 0 < t < \infty \ , \tag{9}$$

will respectively occur, and the boundary condition of Dirichlet's problem for the function $v(x,y)$

$$v(x,y) = 0 \ , \quad (x,y) \in S \ , \tag{10}$$

will take the place of the boundary condition (7).

When $\lambda = 0$ there is no solution which is different from zero and continuous in $G \cup S$ for the problem (8)–(10). The value λ for which the problem (8)–(10) has a nontrivial real solution $v(x,y)$ is called a *eigenvalue* or a *characteristic value*, and $v(x,y)$ the *eigenfunction* corresponding to λ of the problem.

It is easy to see that when the boundary S is piecewise smooth the characteristic value of the problem (8)–(10) is positive.

In fact, for a solution $v(x,y)$ of the equations (8) which is different from a constant in the domain G (we are also interested in this kind of solution) the equality

$$\left(\frac{\partial v}{\partial x}\right)^2 + \left(\frac{\partial v}{\partial y}\right)^2 = \frac{\partial}{\partial x}\left(v\frac{\partial v}{\partial x}\right) + \frac{\partial}{\partial y}\left(v\frac{\partial v}{\partial y}\right) + \lambda v^2$$

holds.

By integrating this equality over the domain G and applying the formula (G-O), we obtain

$$\int_G (v_x^2 + v_y^2)dxdy = \int_S v\frac{dv}{d\nu}ds + \lambda \int_G v^2 dxdy = \lambda \int_G v^2 dxdy \tag{11}$$

on account of (10).

Since v is a real function and different from a constant, we obtain the conclusion that $\lambda > 0$ by the equality (11).

By using the notation $\lambda = \mu^2$, where μ is a real constant, the general solution of the ordinary differential equation (9) can be written down in the form

$$w(t) = c_1 \cos \mu t + c_2 \sin \mu t , \tag{12}$$

where c_1 and c_2 are arbitrary real constants. Evidently, it is sufficient to limit μ to positive values.

Under further general assumptions concerning the domain G the following assertion is valid: The set consisting of all eigenvalues of the problem (8)–(10) is countable:

$$\mu_1^2 \leq \mu_2^2 \leq \dots , \quad \lim_{n \to \infty} \mu_n^2 = \infty . \tag{13}$$

Moreover, the set of the linearly independent eigenfunctions

$$v_1(x, y), v_2(x, y), \dots \tag{14}$$

is also countable, and the equality mark in (13) means that there may exist some kind of eigenvalues which correspond to several linearly independent eigenfunctions. These eigenvalues are said to be *multiple*.

The stated assertion can be verified on the basis of the spectral theory of linear integral operators.

Indeed, writing equation (8) in the form

$$\Delta v(x, y) = f(x, y) , \quad f(x, y) = -\lambda v(x, y) ,$$

and making use of the formula (Poisson's integral)

$$v(x, y) = -\frac{1}{2\pi} \int_G G(x, y; \xi, \eta) f(\xi, \eta) d\xi d\eta ,$$

which gives a solution of the homogeneous Dirichlet problem (10) for this equation [cf. formula (91), Chapter III], we can see that the problem (8)–(10) is equivalent to the linear homogeneous Fredholm integral equation of the second kind with a spectral parameter λ,

$$v(x, y) - \lambda \int_G k(x, y; \xi, \eta) v(\xi, \eta) d\xi d\eta = 0 ,$$

where the kernel of the integral operator on the left-hand side is given by the formula

$$k(x,y;\xi\eta) = \frac{1}{2\pi} G(x,y;\xi,\eta) ,$$

where $G(x,y;\xi,\eta)$ is the Green function of the Dirichlet problem for harmonic functions in the domain G. The kernel $k(x,y;\xi\eta)$ is symmetric with respect to points (x,y) and (ξ,η), positive and has a weak (logarithmic) singularity when $(x,y) = (\xi,\eta)$. As it was well-known that the spectrum of such kind of operators is positive, infinite, discrete and has a unique limit point $\lambda = \infty$. Besides, if S is smooth enough then the minimum element λ_1 of the spectrum (i.e. the first eigenvalue of the problem (8)–(10)) is the minimum of the numbers

$$\int_G \omega(x,y)dxdy \int_G k(x,y;\xi,\eta)\omega(\xi,\eta)d\xi d\eta ,$$

where ω is an arbitrary function in the class $\mathbf{C}^{1,0}(G \cup S)$, satisfying the condition

$$\int_G dxdy \left[\int_S k(x,y;\xi,\eta)\omega(\xi,\eta)d\xi d\eta \right]^2 = 1 .$$

The eigenvalues $\lambda_n, n = 2, \ldots$, can be obtained by repeating the assertion stated here under the supplementary condition

$$\int_G \omega(x,y)\omega_k(x,y)dxdy = 0 , \quad k = 1, \ldots , n-1 .$$

Introducing the notation

$$v^*(x,y) = \int_G k(x,y;\xi,\eta)\omega(\xi,\eta)d\xi d\eta$$

and considering the obvious equality $\Delta x^* = -\omega(x,y), (x,y) \in G$, $v^*(x,y) = 0, (x,y) \in S$ it follows that

$$\int_G dxdy \int_G k(x,y;\xi,\eta)\omega(x,y)\omega(\xi,\eta)d\xi d\eta = -\int_G v^*(x,y)\Delta v^*(x,y)dxdy$$
$$= \int_G ({v_x^*}^2 + {v_y^*}^2)dxdy , \quad \int_G {v^*}^2(x,y)dxdy = 1 ,$$

i.e., λ_1 is the minimum of numbers which are the values of the Dirichlet integral

$$D(v^*) = \int_G \left(v_x^{*^2} + v_y^{*^2}\right) dxdy$$

of the functions v^{*^2} and $v^*(x,y)$ which satisfy the conditions

$$\int_G v^{*^2}(x,y)dxdy = 1 \text{ and } v^*(x,y) = 0 \ , \quad (x,y) \in S \ .$$

Moreover, this minimum is realized by the corresponding eigenfunction $v_1(x,y)$ of λ_1 as follows,

$$\lambda_1 = D(v_1) \ , \quad \int_G v_1^2 dxdy = 1 \ , \quad v_1(x,y) = 0 \ , \quad (x,y) \in S \ .$$

The subsequent eigenvalues and eigenfunctions can be obtained similarly with the only difference that the function $v^*(x,y)$ needs to subject to the supplementary conditions

$$\int_G v^*(x,y)v_k(x,y)dxdy = 0 \ , \quad k = 1, \ldots, n-1 \ .$$

The characters just mentioned of the eigenvalues and the eigenfunctions are usually called the extreme ones. They can also be established by the variational method which we will mention in the following section. However, a more convenient and intuitive method has been used for the cases of membranes with concrete shapes (for example, round, rectangular etc.) to justify applying the method of separation of variables to the problem (5), (6) and (7).

It is not difficult to see that the eigenfunctions v_k and v_m which correspond to the eigenvalues λ_k and λ_m are orthogonal when $\lambda_k \neq \lambda_m$, i.e.,

$$\int_G v_k v_m dxdy = 0 \ . \tag{15}$$

In fact, as a result of integrating over the domain G of the obvious equality

$$\frac{\partial}{\partial x}\left(v_k \frac{\partial v_m}{\partial x} - v_m \frac{\partial v_k}{\partial x}\right) + \frac{\partial}{\partial y}\left(v_k \frac{\partial v_m}{\partial y} - v_m \frac{\partial v_k}{\partial y}\right) = v_k \Delta v_m - v_m \Delta v_k \ ,$$

applying the formula $(G-O)$ and taking into account the equalities

$$\Delta v_k = -\lambda_k v_k \ , \quad \Delta v_m = -\lambda_m v_m \ , \quad (x,y) \in G \ ,$$
$$v_k(x,y) = v_m(x,y) = 0 \ , \quad (x,y) \in S \ ,$$

we have

$$\int_S \left(v_k \frac{dv_m}{d\nu} - v_m \frac{dv_k}{d\nu} \right) ds = \int_G (v_k \Delta v_m - v_m \Delta v_k) dx dy$$
$$= (\lambda_k - \lambda_m) \int_G v_k v_m dx dy \ ,$$

from which follows (15).

Writing the solution (12), which corresponds to the eigenvalue μ_n^2 of equation (9) in the form $w_n(t) = a_n \cos \mu_n t + b_n \sin \mu_n t$, where a_n and b_n are arbitrary real constants, we shall construct the solutions of equation (5) satisfying the boundary condition (7) by the formula

$$u_n(x,y,t) = v_n(x,y)(a_n \cos \mu_n t + b_n \sin \mu_n t) \ , \quad n = 1, 2, \dots \ . \tag{16}$$

Each of the functions (16) describes the proper (free) oscillations of a membrane.

Below we are going to find a solution $u(x,y,t)$ of the problem (5)–(6)–(7) in the form of a sum of a series

$$u(x,y,t) = \sum_{n=1}^{\infty} v_n(x,y)(a_n \cos \mu_n t + b_n \sin \mu_n t) \ . \tag{17}$$

In case that the series on the right-hand side of (17) is uniformly convergent on $Q \cup \partial Q$ and is differentiable up to the second order term-by-term, the sum $u(x,y,t)$ of this series is obviously the solution of equation (5) satisfying the boundary condition (7), since the function $u_n(x,y,t)$ defined by the formula (16) has these properties. The requirement that $u(x,y,t)$ also satisfy the initial conditions (6) is equivalent to the equalities

$$\sum_{n=1}^{\infty} a_n v_n(x,y) = \varphi(x,y) \ , \quad \sum_{n=1}^{\infty} \mu_n b_n v_n(x,y) = \psi(x,y), \ (x,y) \in G \ . \tag{18}$$

Since the system of the functions (14) is linearly independent, we can suppose that it is orthogonal (though the orthogonality of v_n and v_m for $\lambda_n \neq \lambda_m$ has been proved), that is

$$\int_G v_n(x,y) v_m(x,y) dx dy = \begin{cases} 1 \ , & n = m \\ 0 \ , & n \neq m \ . \end{cases} \tag{19}$$

By virtue of (19) from (18) for determining the coefficients a_n and b_n we obtain the formulae

$$a_n = \int_G \varphi(x,y)v_n(x,y)dxdy\ , \quad b_n = \frac{1}{\mu_n}\int_G \psi(x,y)v_n(x,y)dxdy\ . \tag{20}$$

It can be shown that if $\varphi(x,y)$ and $\psi(x,y)$ have enough smoothness then these functions can be represented in the form of the sums of series (18) where the coefficients a_n, b_n are given by the formulae (20), the series on the right-hand side of (17) converging uniformly and its sum $u(x,y,t)$ being a regular solution for the problem (5)–(6)–(7) in the semicylinder Q.

4.1.4. It can be verified that when the boundary S of the domain G is a sufficiently smooth curve, then it is not difficult to show the uniqueness of the solution for the problem (5)–(6)–(7) and the solution itself also has the needed smoothness in $Q \cup \partial Q$.

In fact, let Q_t be the part between planes $\tau = 0$ and $\tau = t$ of the cylindrical domain $Q = \{(\xi,\eta) \in G, 0 < \tau < \infty\}$, and $u(\xi,\eta,\tau)$ is a solution of equation (5) which is regular in Q and satisfies the homogeneous conditions

$$\begin{gathered} u(\xi,\eta,0) = 0\ , \quad u_\tau(\xi,\eta,0) = 0\ , \quad (\xi,\eta) \in G\ , \\ u(\xi,\eta,\tau) = 0\ , \quad (\xi,\eta) \in S\ , \quad \tau \geq 0\ . \end{gathered} \tag{21}$$

The uniqueness of solution for the problem (5)–(6)–(7) would be proved if we can show that the problem (5)–(21) has no solution which is different from zero.

For sufficient smoothness of the curve S we suppose that for the domain Q_t the formula (G-O) is applicable.

Integrating over Q_t the identity

$$-2\left[\frac{\partial}{\partial\xi}\left(\frac{\partial u}{\partial\xi} + \frac{\partial u}{\partial\tau}\right) + \frac{\partial}{\partial\eta}\left(\frac{\partial u}{\partial\eta} + \frac{\partial u}{\partial\tau}\right)\right] + \frac{\partial}{\partial\tau}\left[u_\xi^2 + \mu_\eta^2 + u_\tau^2\right] = 0$$

and applying the formula $(G-O)$, we obtain (compare with the formula (78), Chapter II)

$$\int_{\partial Q_t} \left[-2u_\tau u_\xi \xi_\nu - 2u_\tau u_\eta \eta_\nu + (u_\xi^2 + u_\eta^2 + u_\tau^2)\tau_\nu\right] ds = 0\ , \tag{22}$$

where $\xi_\nu, \eta_\nu, \tau_\nu$ are cosines of the outward normal to ∂Q_t at the point (ξ,η,τ).

The equality

$$\int_{G_t} (u_\xi^2 + u_\eta^2 + u_\tau^2) ds = 0 \tag{23}$$

now follows from (22) on account of (21), where G_t is a domain in the plane $\tau = t$, representing the intersection of Q and the plane $\tau = t$. We obtain from (23) that $u_\xi = u_\eta = u_\tau = 0$ in G_t and hence $u_x = u_y = u_t = 0$ at an arbitrary finite point $(x, y, t) \in Q$. Thus, $u(x, y, t) =$ constant. From this and due to the first of the conditions (21) we conclude that $u(x, y, t) = 0$. At the same time the uniqueness of the solution for the problem (5)–(6)–(7) has been proved.

4.1.5. We turn our attention to the example of a *circular membrane* in detail. Without loss of generality we assume that a circular membrane is present in the circle $x^2 + y^2 \leq 1$ in the plane of variables x and y when it is in equilibrium state.

Introducing polar coordinates (r, ϑ) such that $x = r \cos\, \vartheta, y- = r \sin\, \vartheta$ and making use of the expression

$$\Delta = \frac{\partial^2}{\partial \tau^2} + \frac{1}{\tau}\frac{\partial}{\partial \tau} + \frac{1}{\tau^2}\frac{\partial^2}{\partial \vartheta^2}$$

for the Laplacian $\Delta = \frac{\partial^2}{\partial x^2} + \frac{\partial^2}{\partial y^2}$, we can write Helmholtz's equation (8) in the form

$$r^2 \frac{\partial^2 v}{\partial r^2} + r \frac{\partial v}{\partial r} + \frac{\partial^2 v}{\partial \vartheta^2} + \mu^2 \gamma^2 v = 0\ , \quad \mu^2 = \lambda\ . \tag{24}$$

The method of separation of variables is used again to construct a solution of the problem (10)–(24).

The function $v(r, \vartheta) = R(r)\mathrm{H}(\vartheta)$ will be a solution of the equation (24) if the functions $R(r)$ and $\mathrm{H}(\vartheta)$ are solutions of the ordinary differential equations

$$r^2 R''(r) + rR'(r) + (\mu^2 r^2 - \omega) R(r) = 0 \tag{25}$$

and

$$\mathrm{H}''(\vartheta) + \omega \mathrm{H}(\vartheta) = 0\ , \tag{26}$$

where ω is a real constant

$$\omega = -\frac{\mathrm{H}''}{\mathrm{H}} = \frac{r^2 R'' + rR' + \mu^2 r^2 R}{R} = \text{const.}$$

The boundary condition (10) for the function $v(r,\vartheta)$ is equivalent to the condition $R(1)=0$ for the function $R(t)$.

The function $\mathrm{H}(\vartheta)$ clearly must be periodic with period 2π for single-valuedness of the function $\vartheta(r,\vartheta)$, i.e. $\omega=n^2$ where n is an arbitrary integer. Accordingly to this, the general solution of equation (26) is written in the form

$$\mathrm{H}(\vartheta)=\alpha_n\cos\,n\vartheta+\beta_n\sin\,n\vartheta\;, \tag{27}$$

where α_n and β_n are arbitrary real constants.

With the notations $\mu r=\rho$ and $R(r)=R(\frac{\rho}{\mu})\equiv J(\rho)$, the ordinary differential equation (25) for $R(r)$ becomes *Bessel's equation*

$$J''(\rho)+\frac{1}{\rho}J'(\rho)+\left(1-\frac{n^2}{\rho^2}\right)J(\rho)=0 \tag{28}$$

with integer-valued parameter $\omega=n^2$.

Since we are interested in the solution $v(r,\vartheta)$ which is regular in the circle G of equation (24), we should take from the two linearly independent solutions of the equation (28) a solution which is regular for $0\le r\le 1$ i.e. a *Bessel's function* $J_n(\rho)$ with a non-negative integer-valued index n,

$$J_n(\rho)=\sum_{k=0}^{\infty}(-1)^k\frac{\rho^{n+2k}}{2^{n+2k}k!(n+k)!}\;,\qquad n=0,1,\dots\;.$$

The boundary condition in the discussed case evidently becomes the condition $J_n(\mu)=R(1)=0$ for the function $J_n(\rho)$.

As we know, the Bessel's function $J_n(\mu)$ has a countable number of roots which are all real and the unique limit point is a point at infinity of the plane of complex variable μ. Moreover, there are no Bessel's functions with different indexes that have common roots (except for the root $\mu=0$ if $n>0$).

Therefore, the squares of the roots that are different from zero of the function $J_n(\mu)$ are eigenvalues of the problem (8)–(10). We shall denote these roots by $\mu_{n,m}$ and arrange them in an increasing order:

$$\mu_{n,m}\le\mu_{n,m+1}\le\cdots\;,\qquad m=1,2,\dots,n=0,1,\dots\;.$$

The eigenfunctions which correspond to $\mu_{n,m}^2$ are

$$J_n(\mu_{n,m}r)\cos\,n\vartheta\;,\quad J_n(\mu_{n,m}r)\sin\,n\vartheta\;. \tag{29}$$

It follows from (29) that the eigenvalues $\mu_{0,m}^2$ are all prime and, at the same time, the multiplicity of each eigenvalue $\mu_{n,m}^2, n > 0$, is at least equal to two since the linearly independent eigenfunctions

$$J_n(\mu_{n,m}r)\cos\ n\vartheta\ ,\quad J_n(\mu_{n,m}r)\sin\ n\vartheta$$

correspond to them.

The points in G are called ***nodal points*** if anyone of the eigenfunction vanishes there. For $\mu = \mu_{0,m}, m > 1$, all points on the circumference

$$r = \frac{\mu_{0,m-j}}{\mu_{0,m}}\ ,\quad j = 1,\ldots,m-1\ ,\quad m > 1\ ,$$

and only these are nodes, and for $\mu = \mu_{n,m}, n \geq 1, m > 1$ the circumference

$$r = \frac{\mu_{n,m-j}}{\mu_{n,m}}\ ,\quad j = 1,\ldots,m-1\ ,\quad n = 1,2,\ldots$$

and beams

$$\vartheta = \left(k\pi + \frac{\pi}{2}\right)/n\ ,\quad \vartheta = \frac{k\pi}{n}\ ,\quad k = 0,\ldots,n-1\ ,\quad 0 < r < 1\ ,$$

are nodal. It is clear that the points $(r, \vartheta) \in G$ where the functions

$$\alpha J_n(\mu_{n,m}r)\cos\ n\vartheta + \beta J_n(\mu_{n,m}r)\sin\ n\vartheta\ ,$$

with real constants α and β such that $\alpha^2 + \beta^2 \neq 0$, vanish are also nodes.

Rewriting the corresponding eigenfunction of the $\mu_{n,m}$ in the form

$$v_{n,m} = J_n(\mu_{n,m}r)(\alpha_n \cos\ n\vartheta + \beta_n \sin\ n\vartheta)\ ,$$

where α_n and β_n are arbitrary constants, the function (16), which expresses the proper oscillations of a circular membrane with fixed boundary, has the form

$$u_{n,m} = J_n(\mu_{n,m}r)(\alpha_n \cos\ n\vartheta + \beta_n \sin\ n\vartheta)(a_{n,m}\cos\mu_{n,m}t + b_{n,m}\sin\ \mu_{n,m}t)\ ,$$
$$n = 0,1,\ldots,\quad m = 1,2,\ldots\ .$$

This set of solutions for equation (5) makes it possible to construct a solution of the problem (5)–(6)–(7) according the scheme indicated above.

The equation (8) is sometimes called metaharmonic, and its solutions *metaharmonic functions.*

It was proved at the end of the preceding chapter that the solution of the Dirichlet problem for the equation (8) with $\lambda < 0$ is unique. Under this assumption, the existence of solution for this problem will be proved. When $\lambda > 0$, however, the situation becomes considerably more complex. As we have just verified, when $\lambda = \mu^2_{n,m} > 0$ and in the circle $x^2 + y^2 < 1$ say, the homogeneous Dirichlet problem (10) of equation (8) has linearly independent solutions

$$J_n(\mu_{n,m} r) \cos n\vartheta \ , \quad J_n(\mu_{n,m} r) \sin n\vartheta \ .$$

However, the inhomogeneous Dirichlet problem turns out to be not always solvable.

4.1.6. The characteristic method relates to the classical method of constructing solutions for hyperbolic type equations.

Let $Q(\lambda_1, \dots, \lambda_n)$ be the characteristic quadratic form that corresponds the linear partial differential equation

$$Lu = \sum_{i,j=1}^{n} A_{ij}(x) \frac{\partial^2 u}{\partial x_i \partial x_j} + \sum_{i=1}^{n} B_i(x) \frac{\partial u}{\partial x_i} + C(x) u = 0 \ , \tag{30}$$

and the equation

$$\phi(x_1, \dots, x_n) = 0 \tag{31}$$

be a $(n-1)$-dimensional real surface in the space E_n of variables $x_1, \dots, x_n$. This surface is called a *characteristic surface* of equation (30) if the function $\phi(x_1, \dots, x_n)$ satisfies the equality

$$Q\left(\frac{\partial \phi}{\partial x_1}, \dots, \frac{\partial \phi}{\partial x_n}\right) = 0$$

at all points of the surface (31). It follows from this definition that equation (30) has no characteristic surfaces in the elliptic case. It is clear that each cone of the family

$$\sum_{i=1}^{n} (x_i - x_i^0)^2 - (t - t_0)^2 = 0 \ , \quad x_i^0 = \text{const.} \ , \quad t_0 = \text{const.} \ ,$$

is a characteristic surface of the wave equation

$$\frac{\partial^2 u}{\partial t^2} - \sum_{i=1}^{n} \frac{\partial^2 u}{\partial x_i^2} = 0 . \tag{32}$$

The planes defined by the equation

$$\sum_{i=1}^{n} c_i x_i + c_0 t = \text{const.} , \quad c_0^2 = \sum_{i=1}^{n} c_i^2 ,$$

where $c_i, i = 1, \ldots, n$ are arbitrary real constants, are also contained in the set of the characteristic surfaces for equation (32).

By S we denote a sufficiently smooth n-dimensional surface described in the form

$$\psi(x, t) = 0$$

in the space E_{n+1} of variables $x_1, \ldots, x_n, t$. We assume that on S

$$\sum_{i=1}^{n} \left(\frac{\partial \psi}{\partial x_i} \right)^2 - \left(\frac{\partial \psi}{\partial t} \right)^2 < 0 \tag{33}$$

The surface S which satisfies the condition (33) is called a *surface of space type* for the equation (32). It can be proved that the statement of the following problem for the equation (32) is well-posed: To find a regular solution $u(x, t)$ of equation (32) under the conditions

$$u(x, t) = \varphi(x, t) , \quad \frac{\partial u(x, t)}{\partial l} = \psi_1(x, t) , \quad (x, t) \in S , \tag{34}$$

where l is a sufficiently smooth $(n + 1)$-dimensional unit vector given on S but not a tangent to the plane S at any of its points, and φ and ψ_1 are also sufficiently smooth functions given on S. The problem (32)–(34) just stated is usually called a *Cauchy's problem.* For $n = 1$ the problem (32)–(34) is well-posed and it is so also when, instead of (33), the curve S, the support of the given (34), satisfies the condition

$$\left(\frac{\partial \psi}{\partial x} \right)^2 - \left(\frac{\partial \psi}{\partial t} \right)^2 \neq 0 , \quad x_1 = x .$$

In section 6 of Chapter II, the plane $S : \psi(x,t) \equiv t - t_0 = 0$, which is a surface of space type for the equation (32), was called a *support of the initial condition* given by (55), and the direction l coincided with the axis t.

For the wave equation (32), the plane $x_n = 0$ is neither a characteristic surface nor a surface of space type if the number of spatial variables n is greater than one. The function $u(x,t)$ defined by the formula

$$u(x,t) = \frac{1}{k^2}\text{sh } kx_n \sin\ kx_{n-1} ,$$

where k is a real number, is obviously a solution of the problem with the form (34) and the given conditions

$$u|_{x_n=0} = 0 , \quad \left.\frac{\partial u}{\partial x_n}\right|_{x_n=0} = \frac{1}{k}\sin\ kx_{n-1}$$

on $x_n = 0$ for the equation (32). But this problem is not well-posed, i.e., it is incorrectly formulated because

$$\lim_{k\to\infty} \left.\frac{\partial u}{\partial x_n}\right|_{x_n=0} = 0 ,$$

and at the same time the solution $u(x,t)$ itself is not bounded as $k \to \infty$ (*Hadamard's example*).

On the basis of the example of the wave equation (32) with one spatial variable $x_1 = x$ it is easy to see that the characteristics cannot be used for supports of the initial condition given by (34). Let S coincide with the line $x - t = 0$ which is the characteristic of this equation, and l coincide with the direction of the axis t. Under these assumptions the condition (34) can be written in the form

$$u(x,x) = \varphi(x) , \quad \frac{\partial u}{\partial t}\bigg| t = 0 = \psi_1(x) , \quad -\infty < x < \infty . \tag{35}$$

Making use of the general representation of the solutions of equation (32),

$$u(x,t) = f_1(x+t) + f_2(x-t) ,$$

where f_1 and f_2 are arbitrary functions in the class $\mathbf{C}^{2,0}$, and requiring that the function $u(x,t)$ satisfy the condition (35), we obtain the equalities

$$f_1(2x) + f_2(0) = \varphi(x) , \quad f_1'(2x) - f_2'(0) = \psi_1(x) , \quad -\infty < x < \infty ,$$

which are the ones that could not be satisfied for arbitrary, even if anyhow smooth, φ_1, and ψ_1.

As for the homogeneous problem

$$u(x,x)=0\ ;\quad \left.\frac{\partial u}{\partial t}\right|_{t-x}=0\ ,$$

the expression

$$u(x,t)=f_2(x-t)$$

is a solution, where $f_2(\xi), -\infty<\xi<\infty$, is an arbitrary real function of the class $\mathbf{C}^{2,0}$ and vanishes at $\xi=0$ along with its first order derivative.

Nevertheless, the so-called *Cauchy's characteristic problem* is well-posed in the case of the wave equation. By K we denote the characteristic cone of the equation (32),

$$K:|x-x_0|^2-(t-t^0)^2=0\ ,\quad t\geq t_0\ ,$$

which is the enveloping family of the planes (they are also characteristic of the equation (32)):

$$c(x-x_0)-|c|(t-t_0)\ ,\quad t_0=\text{const.}\ ,\quad x_0=\text{const.}\ ,$$

where C is a real constant n-dimensional vector. In the simplest variant of the Cauchy's characteristic problem it is required to determine in the cone K a regular solution $u(x,t)$ of the equation (32) from the condition

$$u(x,t)=\varphi(x,t)\ ,\quad (x,t)\in K\ , \tag{36}$$

where φ is a given and sufficiently smooth real function on K.

In the case of $n=1$ the cone K degenerates into characteristic straight lines outgoing from the point (x_0,t_0):

$$\begin{aligned}&L_1:x+t=x_0+t_0\ ,\\&L_2:x-t=x_0-t_0\ ,\quad t\geq t_0\ ,\end{aligned}$$

and the condition (36) becomes

$$u(x,t)|_{L_1}=\varphi_1(t)\ ,\quad u(x,t)|_{L_2}=\varphi_2(t)\ ,\quad t\geq t_0\ , \tag{37}$$

where φ_1 and φ_2 are given real functions in the class $\mathbf{C}^{2,0}$ with $\varphi_1(t_0)=\varphi_2(t_0)$.

The characteristic Cauchy's problem (32)–(36) is well-posed, i.e., it has a unique and stable solution. When $n = 1$ the problem (32)–(37) is said to be *Goursat's problem* and its solution is given by the formula

$$u(x,t) = \varphi_1\left(\frac{t - x + x_0 + t_0}{2}\right) + \varphi_2\left(\frac{t + x - x_0 + t_0}{2}\right) - \varphi_1(t_0) \,.$$

For the hyperbolic type equation

$$\frac{\partial^2 u}{\partial t^2} - Lu = f(x,t) \,, \tag{38}$$

where

$$L \equiv \sum_{i,j=1}^{n} A_{ij}(x,t)\frac{\partial^2}{\partial x_i^2} + \sum_{i=1}^{n} B_i(x,t)\frac{\partial}{\partial x_i} + c(x,t)$$

is a linear, uniformly elliptic, positively definite second order operator in spatial variables $x_1, \dots, x_n$, and $f(x,t)$ is a given real function the surface $S : \psi(x,t) = 0$ is called a surface of space type if the inequality .

$$\sum_{i,j=1}^{n} A_{ij}(x,t)\frac{\partial \psi}{\partial x_i}\frac{\partial \psi}{\partial x_j} - \left(\frac{\partial \psi}{\partial t}\right)^2 < 0 \,.$$

holds at its every point. It can be verified that the problem (34)–(38) is well-posed provided the S is a surface of space type and all the given functions in this problem are sufficiently smooth.

Furthermore, the basic mixed problem (6)–(7)–(38) is also well-posed.

Nonstationary processes are in the main modelled in terms of equations of hyperbolic type, while stationary processes, in particular phenomena within media in statically equilibrium states, are usually described by equations of elliptic type. When applying the method of separation of variables, what we have to do most often is to deal with elliptic equations. Parabolic type equations find applications in studying transport phenomena. It is appropriate here to remember that the system of equations that describes a planar stationary irrotational motion of a viscous compressible medium is elliptic in the subsonic domain, hyperbolic in the supersonic domain, and parabolically degenerates along a sound line.

In Chapter III we have made sure that in the theory of the classical problem of linear elliptic type equations the method of parametrix plays an important role. This method allows us to apply linear integral equations to the investigation of solvability problems for these stated problems. In a certain sense the thermal potential method, which has been successfully applied to investigation of classical problems of parabolic type equations, is a modification of the stated method.

Partial differential equations used in applications are in most cases differential representations of conservation laws which the investigated phenomena obey. These laws are usually formulated in the form of integrals, and functions which the describe the phenomena are solutions of variational problems on minimization of functionals of specific forms. Methods that allow us to get these functions without applying partial differential equations are called *variational methods.* The next section is devoted to some of these kinds of methods.

§4.2 Variational Methods

4.2.1. As we have already noticed in §1.1 of Chapter I, the deflection $u = u(x,y)$ of an elastic membrane in equilibrium state is, under certain assumptions, a harmonic function of variables x and y on a domain G, that is, a regular solution of Laplace's equation

$$\frac{\partial^2 u}{\partial x^2} + \frac{\partial^2 u}{\partial y^2} = 0 \ , \quad (x,y) \in G \ , \tag{39}$$

which this time plays the role of Euler-Lagrange's equation for Dirichlet's integral

$$D(u) = \int_G (u_x^2 + u_y^2)\,dx\,dy \ , \tag{40}$$

representing the potential energy of a membrane in a bending state.

Let $\varphi(x,y)$ be a real function given on $S = \partial G$ and of the class $\mathbf{C}^{0,0}(S)$. The functions $\{u(x,y)\}$ in the class $\mathbf{C}^{0,0}(G \cup S)$ which have piecewise continuous first order derivatives in G, for which the Dirichlet integral (40) exists (i.e., finite), and, moreover, which satisfy the boundary condition

$$u(x,y) = \varphi(x,y) \ , \quad (x,y) \in S \ , \tag{41}$$

are said to be *admissible functions.*

The first boundary problem or Dirichlet's problem, formulated as follows, has been studied in Chapter III: Among the functions that are harmonic in a domain G and the class $\mathbf{C}^{0,0}(G \cup S)$, we have to seek one such that the boundary condition (41) is satisfied. Moreover, under quite general assumption concerning the domain G, it was proved that the solution of this problem exists and is unique and stable.

For the physical problem from the unknown form $u = u(x, y)$ of a membrane situated in equilibrium bending state, it is required that the function $u(x, y)$ should be in the class of admissible functions and in this class it should minimize Dirichlet's integral. The problem of searching a function among the admissible functions to minimize Dirichlet's integral is called the *first variational problem*. It should be emphasized that nothing is said about the second derivatives of the unknown function in the formulation of the first variational problem.

The following statement is true: If the function $\varphi(x, y)$ given on S is such that the class of admissible functions is not empty, then Dirichlet's problem and the first variational problem are equivalent.

Now we will show the validity of this claim under some supplementary assumptions which will be mentioned below.

Let $u(x, y)$ be a solution of the first variational problem. We represent the class of admissible functions in the form of $u(x, y) + \varepsilon h(x, y)$, where ε is an arbitrary constant and $h(x, y)$ an arbitrary function in the class of admissible functions satisfying the boundary condition

$$h(x, y) = 0 \ , \quad (x, y) \in S \ . \tag{42}$$

Evidently,

$$\mathcal{D}(u + \varepsilon h) = D(u) + 2\varepsilon D(u, h) + \varepsilon^2 D(h) \ , \tag{43}$$

where

$$D(u, h) = \int_G (u_x h_x + u_y h_y) dx dy \ . \tag{44}$$

Because $u(x, y)$ is the minimizing function and ε is an arbitrary constant, we obtain from (43)

$$D(u, h) = 0 \ . \tag{45}$$

We shall accept the following supplementary assumptions: the functions $u(x,y), h(x,y)$ and the boundary S of domain G are smooth enough so that the equalities

$$u_x h_x + u_y h_y = (u_x h)_x + (u_y h)_y - h\Delta u$$

and

$$D(u,h) = \int_S h\frac{du}{d\nu}ds - \int_G h\Delta u dx dy \tag{46}$$

hold, where ν is the outward normal to S, and moreover, Δu is continuous in G.

Due to (42) and (45) we have from (46)

$$\int_S h\Delta u \; dx dy = 0 \; .$$

Under the assumption that Δu is a continuous function in G we conclude from this and the arbitrariness of $h(x,y)$ that $\Delta u(x,y) = 0$. Therefore, under the above assumptions the solution of the first variational problem is the solution of Dirichlet's problem (39)–(41).

Now let $u(x,y)$ be a solution of Dirichlet's problem (39)–(41), for which Dirichlet's integral exists. As above, the class of admissible functions can be represented in the form $u(x,y) + \varepsilon h(x,y)$.

In addition, we will assume that for $u(x,y)$ and $h(x,y)$ the identity (46) holds. Due to (39) and (42), the equality (45) follows from this identity and, hence, we obtain

$$D(u) \leq D(u + \varepsilon h)$$

on the basis of (43). But this means that the function $u(x,y)$ minimizes Dirichlet's integral (40), i.e., it is the solution of the first variational problem.

The idea of reducing the Dirichlet boundary value problem for Laplace's equation to the first variational problem for Dirichlet's integral belongs to Riemann. The assertion about the equivalence between these two problems that was stated above, which holds under the assumption that the class of admissible functions is not empty, is well-known as *Dirichlet's principle.*

The equivalence between Dirichlet's problem (39)–(41) and the first variational problem is far from being so always. It can be verified by a simple example that Dirichlet's problem (39)–(41) can have one and only one solution $u(x,y)$ but, at the same time, neither $u(x,y)$ minimizes Dirichlet's integral $D(u)$ nor this integral converges.

Let G be a unit circle $|z|^2 = x^2 + y^2 < 1$ and for the boundary condition (41),

$$u(x,y) = \begin{cases} 1\,, & 0 < \vartheta < \pi \\ 0\,, & \pi < \vartheta < 2\pi\,, \end{cases} \tag{47}$$

where $x + iy = e^{i\vartheta}$ is a point on the circumference $|z| = 1$.

The expression

$$u(x,y) = \frac{1}{\pi}\text{Im}\ \log\ i\frac{1+z}{1-z}\,, \quad |z| < 1\,,$$

is obviously a solution of the problem (39)–(47), where $\log\ i\frac{1+z}{1-z}$ is the branch of this function that is equal to $\frac{i\pi}{2}$ at $z = 0$.

Because of

$$u_x^2 + u_y^2 = \frac{1}{\pi^2}\left|\left(\log\frac{1+z}{1-z}\right)'\right|^2 = \frac{4}{\pi^2}\left|\frac{1}{1-z^2}\right|^2\,,$$

Dirichlet's integral $D(u)$ diverges.

Taking for an example, the function

$$u(x,y) = \text{Re}\ f(z) \equiv \text{Re}\sum_{k=1}^{\infty}\frac{z^{2^{2k}}}{2^k}\,, \quad z = x + iy\,,$$

which is harmonic in the circle $|z| < 1$ and has continuous boundary values

$$u = \sum_{k=1}^{\infty}\frac{1}{2^k}\cos\ 2^{2k}\vartheta\,, \quad 0 \le \vartheta \le 2\pi\,.$$

It can be verified that for $0 < r < 1$ and $\rho e^{i\vartheta} = z$,

$$\begin{aligned}\int_{|z|\le r}(u_x^2 + u_y^2)dxdy &= \int_{|z|\le r} f'(z)\overline{f'(z)}dxdy \\ &= 2\pi\int_0^r\sum_{k=1}^{\infty}2^{2k}\rho^{22^{2k}-1}d\rho = \pi\sum_{k=1}^{\infty}r^{2^{2k+1}}\,,\end{aligned}$$

and, hence, Dirichlet's integral $D(u)$ taken on the circle $|z| \leq 1$ is divergent this time.

A similar situation will appear if the smoothness of the boundary S of domain G is violated.

4.2.2. In the preceding paragraph, the construction of functions that describe the oscillations of an elastic membrane, present in a finite domain G in the plane of variables x and y and situated in the equilibrium state, was carried out by reducing it to a spectral problem formulated as follows: it is required to determine the eigenvalues and the eigenfunctions of Helmholtz's equation

$$\Delta u + \lambda u = 0 \;, \quad (x,y) \in G \;, \quad \lambda = \text{const.} \;, \tag{48}$$

i.e., to find a value λ for which this equation in the domain G has nontrivial solutions that satisfy the homogeneous boundary condition

$$u(x,y) = 0 \;, \quad (x,y) \in S = \partial G \;, \tag{49}$$

and to construct them.

Let us introduce for discussion the functional

$$J(u) = \frac{D(u)}{\mathcal{H}(u)} \;, \tag{50}$$

where $D(u)$ is Dirichlet's integral (40) and

$$\mathcal{H}(u) = \int_G u^2(x,y)dxdy \;.$$

It will be assumed that for the functional (50) the real functions which are different from zero, satisfying the boundary condition (49), being continuous on $G \cup S$, having piecewise continuous first order derivatives in G, and for which Dirichlet's integral (40) exists, are admissible.

By the *second variational problem* we mean the problem to find in the class of admissible functions, minimum of the functional (50), and to construct the corresponding minimizing function.

We now show that under certain supplementary assumptions and when a solution of the second variational problem exists, $\lambda = J(u)$ is the smallest eigenvalue of the problem (48)–(49), if $u(x,y)$ is the minimizing function, and $u(x,y)$ is the eigenfunction corresponding to the λ for the same problem.

In fact, let the function $u(x,y)$ minimizes $J(u)$ and

$$\lambda = J(u) = \frac{D(u)}{\mathcal{H}(u)} . \tag{51}$$

For the class of admissible functions $u(x,y) + \varepsilon h(x,y)$, where ε is an arbitrary constant and $h(x,y)$ an arbitrary admissible function we have $J(u + \varepsilon h) - J(u) \geq 0$, or

$$2\varepsilon[D(u,h)\mathcal{H}(u) - \mathcal{H}(u,h)D(u)] - \varepsilon^2[D(h)\mathcal{H}(u) - D(u)\mathcal{H}(h)] \geq 0 ,$$

where

$$\mathcal{H}(u,h) = \int_G uh dx dy .$$

This estimate is possible if

$$D(u,h)\mathcal{H}(u) - \mathcal{H}(u,h)D(u) = 0 .$$

In turn, on account of (51) we obtain from this

$$D(u,h) - \lambda\mathcal{H}(u,h) = 0 . \tag{52}$$

Assuming certain conditions of the smoothness of the functions $u(x,y)$ and $h(x,y)$, and of the boundary S of the domain G so that the formula (46) applies, the identity (52) can be rewritten in the form

$$\mathcal{H}(\Delta u + \lambda u, h) = 0 . \tag{53}$$

Then it follows from this identity that $\Delta u + \lambda u = 0$. That is, $u(x,y)$ is the eigenfunction which corresponds to the eigenvalue λ for the problem (48)–(49).

Let λ^* be an eigenvalue that is different from zero, and $u^*(x,y)$ the corresponding eigenfunction for the problem (48)–(49), for which the formula (46) holds. We have on the basis of this formula

$$\mathcal{H}(\Delta u^* + \lambda^* u^*, u^*) = -D(u^*) + \lambda^*\mathcal{H}(u^*) = 0$$

and, hence,

$$\lambda^* = \frac{D(u^*)}{\mathcal{H}(u^*)} \geq \frac{D(u)}{\mathcal{H}(u)} = \lambda .$$

That is, among the eigenvalues of the problem (48)–(49) λ is the smallest.

The inverse assertion is also true: if λ is the smallest eigenvalue of the spectral problem (48)–(49), then its corresponding eigenfunction $u(x, y)$ in the class of admissible functions for the functional $J(u)$ is the solution of the second variational problem, and furthermore $\lambda = \min J(u)$.

Indeed, when $h = u$ the equality (51) can be obtained from (53) and the identity (52), which is equivalent to (53), becomes the identity $D(u, h)\mathcal{H}(u) - H(u, h)D(u) = 0$ on account of (51) which guarantees the estimate $J(u+\varepsilon h) - J(u) \geq 0$, so that the validity of the stated inverse assertion is proved.

We use the notations $\lambda = \lambda_1$ and $u = u_1$ for the smallest eigenvalue and its corresponding eigenfunction for the problem (48)–(49).

Because $cu_1(x, y)$, where c is an arbitrary real non-zero constant, are also eigenfunctions corresponding to λ_1 in addition to $u_1(x, y)$, we can assume without loss of generality that

$$\mathcal{H}(u_1) = 1 \ , \quad D(u_1) = \lambda_1 \tag{54}$$

We now prove that the next eigenvalue λ_2, after λ_1, and its corresponding eigenfunction $u_2(x, y)$ of the problem (48)–(49) can be obtained as a result of solving the variational problem

$$\min J(u) = J(u_2) = \lambda_2$$

in the class of admissible functions that satisfy the condition

$$\mathcal{H}(u, u_1) = 0 \ . \tag{55}$$

Assuming that the solution of the stated variational problem exists, we can repeat the reasoning used above for deriving the identity (52) and conclude that

$$D(u_2, \xi) - \lambda_2 \mathcal{H}(u_2, \xi) = 0 \tag{56}$$

for an arbitrary admissible function ξ that satisfies the condition (55). It is easy to see that the identity (56) holds for any admissible function $\eta(x, y)$ which must not satisfy the condition (55). In fact, from the condition (55) and the identity (52) in which $\lambda = \lambda_1, u = u_1$ and $h = u_2$ we have

$$\mathcal{H}(u_2, u_1) = 0 \ , \quad D(u_2, u_1) = 0 \ . \tag{57}$$

On the basis of the first equality of (54), it is clear that for any admissible function $\eta(x,y)$ the function

$$\xi_1(x,y) = \eta(x,y) - H(\eta, u_1)u_1 \tag{58}$$

satisfies the condition (55) and hence

$$D(u_2, \xi_1) - \lambda_2 \mathcal{H}(u_2, \xi_1) = 0 \ . \tag{59}$$

Replacing the expression $\xi_1(x,y)$ in (59) by (58) and considering the equality (57), we conclude that

$$D(u_2, \eta) - \lambda_2 \mathcal{H}(u_2, \eta) = 0 \ . \tag{60}$$

Since the identity (60) is true for an arbitrary admissible function, under the requirement that these functions and the boundary S of the domain G are sufficiently smooth we can attain from (60)

$$H(\Delta u_2 + \lambda_2 u_2, h) = 0 \ ,$$

as we have done to derive the identity (53), i.e., $\Delta u_2 + \lambda_2 u_2 = 0$.

It can be verified similarly that if for any positive integer n

$$\lambda_n = J(u_n) = \min J(u) \tag{61}$$

in the class of admissible functions that satisfy the conditions

$$\mathcal{H}(u, u_i) = 0 \ , \quad i = 1, \ldots, n-1 \ , \tag{62}$$

then λ_n and $u_n(x,y)$ are an eigenvalue and its corresponding eigenfunction of the problem (48)–(49) and, moreover,

$$D(u_i) = \lambda_i \ , \ \mathcal{H}(u_i) = 1 \ , \ \mathcal{H}(u_i, u_k) = 0 \ ; \quad i \neq k \ , \ i, k = 1, \ldots, n \ . \tag{63}$$

Because the class of admissible functions for the variational problem (61) is narrower for $n = k+1$ than for $n = k$, it is evident that

$$\lambda_1 \leq \lambda_2 \leq \ldots \leq \lambda_n \leq \ldots \ .$$

Under some quite general assumptions on the smoothness of the boundary S of the domain G the following assertions are true:

(1) for any positive integer n the variational problem (61)–(62) has solutions; that is, in the class of admissible functions $J(u)$ has its minimum $\lambda_n = J(u_n)$ and the functions u_n satisfy the condition (63);

(2) the value λ_n and the function $u_n(x, y)$ are a solution of the eigenvalue problem (48)–(49);

(3) $\lim_{n\to\infty} \lambda_n = \infty$;

(4) the system of eigenfunctions

$$u_1(x, y) , \quad u_2(x, y), \dots \tag{64}$$

is complete in the sense that for the arbitrary function $f(x, y) \in \mathbf{C}^{0,0}(G \cup S)$ and a given arbitrary number $\varepsilon > 0$ there exists a linear form $L_n = \sum_{k=1}^{n} c_k u_k(x, y)$ with real coefficients c_k such that

$$\mathcal{H}(f - L_n) < \varepsilon ,$$

and, moreover, when n is fixed, the minimal value of $H(f - L_n)$ can be reached when c_k are the *Fourier coefficients* of the function f with respect to the system (64), i.e.,

$$c_k = a_k = \int_G f(x, y) u_k(x, y) dx dy , \quad k = 1, \dots, n ; \tag{65}$$

(5) for the Fourier coefficients (65) the equality

$$\sum_{k=1}^{\infty} a_k^2 = H(f)$$

holds;

(6) if $f(x, y)$ belongs to the class of admissible functions for the functional J, then it can be represented in the form of an absolutely and uniformly convergent the Fourier series:

$$f(x, y) = \sum_{k=1}^{\infty} a_k u_k(x, y) .$$

We are not going to dwell on the proofs of these assertions here.

4.2.3. Some of the methods for constructing the solutions of the variational problems mentioned above are associated with the names of Gauss and Riemann. These methods are usually called *variational methods.* Historically, certain difficulties in the development of the variational methods have been caused by the possibility that the minimizing function might not exist in the class of admissible functions. For example, in the class of admissible functions which are in the class $\mathbf{C}^{0,0}(0 \leq x \leq 1)$ and satisfy the conditions

$$u(0) = 1 , \quad u(1) = 1 , \tag{66}$$

a real function $u(x)$ is such that the functional

$$J_*(u) = \int_0^1 u^2(x)dx \tag{67}$$

is minimized does not exist.

Since for any positive integer n the function $u_n(x) = x^n$ is admissible and

$$\lim_{n\to\infty} J_*(u_n) = \lim_{n\to\infty} \frac{1}{2n+1} = 0 ,$$

the lower bound of the functional (67) is equal to zero. Thus, a real function in $\mathbf{C}^{0,0}(0 \leq x \leq 1)$ which minimizes the functional (67) should be identically equal to zero. But it cannot be admissible because it does not satisfy the second one of the conditions (66).

When the class of admissible functions $\{u\}$ is not empty, the set of the values of their Dirichlet's integrals $D(u)$ has a lower bound d. It is obvious that there exists a sequence of admissible functions $u_n(x, y), n = 1, 2, \ldots$, such that

$$\lim_{n\to\infty} D(u_n) = d \tag{68}$$

even if we do not know whether this bound of admissible functions can be achieved.

The sequence $\{u_n\}, n = 1, 2, \ldots$, for which the equality (68) holds, is said to be a *minimizing sequence.*

The same thing can be said of the functional $J(u)$.

The existence of the minimizing sequence does not mean the existence of solution for the considered variational problem. The following problems should be the object for further investigation: (1) How to construct the minimizing sequence? (2) Does it converge? (3) Is its limit $u = \lim_{n\to\infty} u_n$ an admissible function?

A detailed investigation of this problem needs to introduce for examination some functional spaces where, particularly, the terms of the minimizing sequence are: elements of them. Having established the convergence of the minimizing sequence in the metric of these spaces, it is desirable to either show that the obtained limit is a solution of the problem stated above or generalize the concept of the solution itself in a reasonable way. In the meantime it is important to establish that a solution for the variational problem is either a solution of the corresponding boundary problem in the ordinary sense or that in a certain generalized sense.

In calculus of variations there exist some methods for solving variational problems, in which partial differential equations are not used. Such methods in connection with problems for partial differential equations are usually called *variational* or *direct methods*. It is important to note that some of the direct methods allow constructing approximate solutions for the considered problems. We will mention two such methods below. The first is named *Ritz's method*.

4.2.4. The essence of Ritz's method is as follows. Consider the problem of minimizing a functional $\phi(u)$. By $\{v_n\}, n = 1, 2, \ldots$, we denote a complete system of admissible functions for the functional $\phi(u)$, and make up a sequence

$$u_n = \sum_{k=1}^{n} c_k v_k \ , \quad n = 1, 2, \ldots \ , \tag{69}$$

where c_k are arbitrary constants at the moment.

We will determine the coefficients $c_k, k = 1, 2, \ldots , n$, such that the expression $\varphi_n = \phi(u_n)$, as function of $c_1, \ldots , c_n$, was minimal.

It has been successfully proved by Ritz that for some classes of functionals the sequence (69), constructed in the indicated way, is a minimizing sequence that converges, and its limit is a solution of the problem under consideration.

As an example, consider the second variational problem of minimizing the functional $J(u)$, when the domain G is be the square $\{0 < x < \pi, 0 < y < \pi\}$

and let

$$\mathcal{H}(u) = 1 \ . \tag{70}$$

The system of functions

$$\{\sin\ kx \sin\ ly\} \ , \quad k, l = 1, 2, \dots \ ,$$

is used as the complete system stated above.

Let

$$u_{mn} = \sum_{k=1}^{m} \sum_{l=1}^{n} c_{kl} \sin\ kx \sin\ ly \ , \quad m, n = 1, 2, \dots \ .$$

The functions u_{mn} obviously satisfy the condition (49) and are admissible for the functional $J(u)$. In addition,

$$D(u_{mn}) = d_{mn} = \frac{\pi^2}{4} \sum_{k=1}^{m} \sum_{l=1}^{n} c_{kl}^2 (k^2 + l^2) \ , \tag{71}$$

$$\mathcal{H}(u_{mn}) = h_{mn} = \frac{\pi^2}{4} \sum_{k=1}^{n} \sum_{l=1}^{n} c_{kl}^2 \ . \tag{72}$$

Now what we have to do along Ritz's scheme is to find the minimum of the expression (71) under the condition

$$\sum_{k=1}^{m} \sum_{l=1}^{n} c_{kl}^2 = \frac{4}{\pi^2} \tag{73}$$

by virtue of (70) and (72).

Solving this problem to minimize (71) and (73) we find that for any m and n all the c_{kl}, except c_{11} are equal to zero, and

$$c_{11} = \frac{2}{\pi} \ , \quad d_{mn} = 2 \ ,$$

that is,

$$\lim_{\substack{m \to \infty \\ n \to \infty}} u_{mn} = u(x, y) = \frac{2}{\pi} \sin\ x \sin\ y \ ,$$

$$\lim_{\substack{m \to \infty \\ n \to \infty}} h_{mn} = 1 \ , \quad \lim_{\substack{m \to \infty \\ n \to \infty}} d_{mn} = D(u) = 2 \ .$$

Ritz's method allows constructing approximate solutions of the eigenvalue problem (48)–(49). In fact, we can take the function

$$u_n(x,y) = \sum_{k=1}^{n} c_k v_k(x,y) \tag{74}$$

from the sequence (69) for an approximate solution of the second variational problem on minimizing the functional $J(u)$ under the condition (70), where the coefficients $c_k, k = 1, 2, \ldots$, are determined by solving the problem on conditional minimum:

$$d_n(c_1, \ldots, c_n) = D(u_n) = \min , \quad h_n(c_1, \ldots, c_n) = H(u_n) = 1 .$$

By the method stated here, the constructed function u_n is taken for an approximate expression of an eigenfunction, and the number λ_n given by formula $\lambda_n = D(u_n)$ for the corresponding eigenvalue of the problem (48)–(49).

4.2.5. The *Bubnov-Galerkin method* has been used successfully for constructing an approximate solution of the same problem. In this method, for an approximate solution of the problem (48)–(49) we take a function $u_n(x,y)$ given by the formula (74), where the coefficients $c_k, k = 1, \ldots, n$ are determined by the equalities (cf. equation (53))

$$\mathcal{H}(\Delta u_n + \lambda u_n \ , \ v_m) = 0 \ , \quad m = 1, \ldots, n,$$

or just the same, by

$$\sum_{k=1}^{n} \mathcal{H}(\Delta v_k + \lambda v_k \ , \ v_m) c_k = 0 \ , \quad m = 1, \ldots, n \ . \tag{75}$$

The equality (75) is a homogeneous system of linear algebraic equations where the number of unknown functions equals the number of the equations. We know from linear algebra that the system (75) has nontrivial solutions iff λ satisfies the equation

$$\det \|\mathcal{H}(\Delta v_k + \lambda v_k \ , \ v_m)\| = 0 \ . \tag{76}$$

The values of λ determined by equation (76) are used as approximate expressions of the eigenvalues of the problem (48)–(49). As already indicated, the corresponding approximate expressions of eigenfunctions are given by the formula (74), where $c_k, k = 1, \ldots, n$, are nontrivial solutions of the system (75).

To estimate the error of using the approximation for the exact solution when applying the Bubnov–Galerkin method, we shall meet with the same difficulties as those encountered, when we apply Ritz's method. As to how we can surmount them we can say something only when we are actually solving concrete problems.

Compared with Ritz's method, the Bubnov–Galerkin method is preferable, first of all, because solving equations (75)–(76) is simpler than solving the problem on conditional minimum as in Ritz's method.

§4.3 Generalized Solutions for Classical Problems of Partial Differential Equations

4.3.1. The concept of Dirichlet's integral has been extended to functions given in a multi-dimensional domain. Let G be a bounded domain in the n-dimensional Euclidean space E_n of points x with Cartesian orthogonal coordinates $x_1, \ldots, x_n, n > 2$, with boundary $S = \partial G$ which is a $(n-1)$-dimensional surface, and $u(x)$ be a real function given in G. The functional

$$D(u) = \int_G \nabla u \cdot \nabla u dx \ , \tag{77}$$

where $\nabla = \text{grad}, \nabla u \cdot \nabla u = \sum_{i=1}^{n} u_{x_i}^2$ and dx is a volume element, is called *Dirichlet's integral.*

A function $u(x)$ is said to be an admissible function if it is continuous in $G \cup S$, has piecewise continuous first order derivatives, coincides with a given continuous function $g(x)$.

$$u(x) = g(x) \ , \quad x \in S \ , \tag{78}$$

and for which Dirichlet integral (77) exists. The problem of determining a function $u(x)$ for which the Dirichlet's integral is minimal among the admissible functions is called the first variational problem.

Just as in the case of $n = 2$, it can be proved by the direct method of calculus of variations that for $n > 2$, the first variational problem also has a unique solution if the boundary S of the domain G and the function $g(x)$ given on it are such that the class of admissible functions is not empty. If $u(x)$ is an admissible function, then the class of all admissible functions can be expressed in the form of $\{u(x) + \varepsilon h(x)\}$, where $h(x)$ is an arbitrary admissible function which vanishes on S and ε is an arbitrary real constant.

From the equality

$$D(u + \varepsilon h) = D(u) + 2\varepsilon D(u, h) + \varepsilon^2 D(h) , \tag{79}$$

it follows that $u(x)$ is the solution for the first variational problem iff the equality

$$D(u, h) = 0 \tag{80}$$

holds, where

$$D(u, h) = \int_G \nabla u \cdot \nabla h dx .$$

In fact, the sufficiency of the condition (80) is obvious because if it is true then we have the estimate

$$D(u) \leq D(u + \varepsilon h)$$

from (79), and this means that $u(x)$ is a minimizing function. To show the necessity of the condition (80) it is enough to note that if $D(u, h) \neq 0$ we can choose the sign of ε so that the inequality $\varepsilon D(u, h) < 0$, and hence

$$D(u) > D(u + \varepsilon h)$$

by (79); that is, $u(x)$ is not a minimizing function.

The Euler-Lagrange equation of the functional (77) is Laplace's equation $\Delta u = 0$. It is clear that when $n > 2$ there is an equivalence between the first variational problem of Dirichlet's integral and Dirichlet's problem (78) for harmonic functions (Dirichlet's principle).

If the boundary S of the domain G and the functions $u(x)$ and $h(x)$ given in G are smooth enough, and the boundary conditions $u(x) = 0, h(x) = 0, x \in S$, are satisfied, the following equalities hold:

$$\nabla u \cdot \nabla h = \nabla(h \nabla u) - h \Delta u = \nabla(u \nabla h) - u \Delta h ,$$
$$D(u, h) = -\mathcal{H}(h, \Delta u) = -\mathcal{H}(u, \Delta h) .$$

The latter is still true if, instead of the condition $u(x) = 0, x \in S$, we have $h(x) = 0$ on the boundary band of S.

When we do not know if there exist the partial derivatives $u_{x_i x_i}, i = 1, \ldots, n$, while we do know that the equality

$$\int_G vhdx = \int_G u\Delta hdx$$

holds for a completely definite function v and any function h which is sufficiently smooth in G and vanishes on the boundary band of S, we naturally consider that the Laplacian of the function $u(x)$ exists in a generalized sense, i.e., $v = \Delta u$. This case suggests that $u(x)$, which satisfies the identity

$$D(u,h) + \mathcal{H}(f,h) = 0$$

for any function $h(x)$ which is smooth enough in G and vanishes on S, is a generalized solution of Dirichlet's problem $u(x) = 0, x \in S$, for Poisson's equation $\Delta u = f(x), x \in G$.

The demonstration of Dirichlet's principle and the concept of the generalized Laplacian that we have just stated comes from Riemann.

4.3.2. One version of generalized derivatives is introduced as follows.

Assume that G is a domain in the space E_n but there is no assumption concerning the boundedness of G and the structure of its boundary. A function $\varphi(x), (x = (x_1, \ldots, x_n))$ given in G is said to be *compactly supported* if it is infinitely differentiable and identically equal to zero outside some compact set $K \subset G$. The compact K is called the support of the function $\varphi(x)$. As every compactly supported function $\varphi(x)$ has its own support K, we sometimes write $K = \operatorname{supp} \varphi$.

For the ordinary m-th order derivatives of the function $\varphi(x)$ we use the notation

$$D^m \varphi = \frac{\partial^m \varphi}{\partial x_1^{m} \ldots \partial x_n^{m_n}}, \quad m = \sum_{j=1}^n m_j .$$

We consider two functions $u(x)$ and $v(x)$ which are given in G and are locally integrable (i.e. integrable in every subdomain δ for which $\delta \cup \partial\delta \subset G$).

Definition: The function $v(x)$ is said to be a *generalized* m-th order derivative of the function $u(x)$ if the identity

$$\int_G uD^m \varphi dx = (-1)^m \int_G v\varphi dx \tag{81}$$

holds for every compactly supported function $\varphi(x)$ with compact support in G.

If $v(x)$ is a generalized m-th order derivative of the function $u(x)$, then we write

$$v = D^m u .$$

If in addition to (81) we also have the identity

$$\int_G uD^m \varphi dx = (-1)^m \int_G v_1 \varphi dx ,$$

where u and φ are the same functions as in (81) and v_1 is a function locally integrable in G, then we have

$$\int_G (v - v_1)\varphi dx = 0 .$$

It follows from this that $v(x) = v_1(x)$ are a.e. in G, i.e., a generalized derivative is uniquely defined if it does exist.

The existence of the generalized first order derivative v for the function $u(x)$ obliges the last function to be at least absolutely continuous. To verify this, we confine ourselves to observing the case of $n = 1$ and use the notations

$$G : \{a < x < b\} , \quad x = x_1 , \quad u(x) = u(x_1) , \quad v(x) = v(x_1) , \quad \varphi(x) = \varphi(x_1) .$$

Since $v(x) = D'u(x)$ by definition, we should have

$$\int_a^b u(x)D'\varphi(x)dx = -\int_a^b v(x)\varphi(x)dx \tag{82}$$

for every compactly supported function $\varphi(x)$ with a compact support in the interval $a < x < b$.

Introducing the function

$$w(x) = \int_a^x v(t)dt ,$$

the right-hand side of the formula (82) can be rewritten in the form

$$-\int_a^b v(x)\varphi(x)dx = \int_a^b w(x)D'\varphi(x)dx , \tag{83}$$

as a result of integration by parts and the consideration that $\varphi(x)$ is compactly supported.

From (82) and (83) we obtain

$$\int_a^b [u(x) - w(x)]D'\varphi(x) = 0 . \tag{84}$$

Due to (84) the function $u(x)$ is expressed by the formula

$$u(x) = w(x) + c$$

a.e. on the interval (a, b), where c is an arbitrary constant. But $w(x) + c$ being an absolutely continuous function as an indefinite Lebesgue integral, the function $u(x)$ must be the same.

For the generalized derivatives some of the properties of the usual (classical) derivatives are still true, but not all. As we know, the existence of the usual (ordinary) m-th order derivatives of a function $u(x)$ guarantees the existence of all derivatives of order less than m of the same function. This property of the usual derivatives may not hold for generalized derivatives. We can see this for ourselves from the example of a function $u(x_1, x_2) = f_1(x_1) + f_2(x_2)$, where $f_1(x_1)$ and $f_2(x_2)$ are arbitrary locally integrable functions; which have no generalized derivatives. It is easy to see that the function $u(x_1, x_2)$ has generalized mixed derivatives $D^2u \equiv \frac{\partial^2 u}{\partial x_1 \partial x_2} \equiv 0$ a.e. in G. In fact, we assume that the compact support of the function $u(x_1, x_2)$ is in the rectangle

$$\{a_1 < x_1 < b_1 , \quad a_2 < x_2 < b_2\} .$$

The correctness of the stated assertion then follows from the obvious equality

$$\int_G u \frac{\partial^2 \varphi}{\partial x_1 \partial x_2} dx_1 dx_2 = \int_{a_1}^{b_1} dx_1 \int_{a_2}^{b_2} [f_1(x_1) + f_2(x_2)] \frac{\partial^2 \varphi}{\partial x_1 \partial x_2} dx_2$$

$$= \int_{a_1}^{b_1} f_1(x_1)dx_1 \int_{a_2}^{b_2} \frac{\partial^2 \varphi}{\partial x_1 \partial x_2} dx_2 + \int_{a_2}^{b_2} f_2(x_2)dx_2 \int_{a_1}^{b_1} \frac{\partial^2 \varphi}{\partial x_1 \partial x_2} dx_1 = 0 .$$

When the function $u(x)$ has a locally integrable ordinary derivative $D^m u$, this derivative must coincide with the generalized derivative. To see this we integrate by parts the left-hand side of the formula (81) and take into consideration the compactly-supportedness of the function φ, then

$$\int_G uD^m\varphi dx = (-1)^m \int_G D^m u \cdot \varphi dx \ .$$

4.3.3. Let $L_2(G)$ be the *Hilbert space* of the functions given in G (i.e. functions square-integrable over G). By W_2^1 we denote the set of those functions in $\mathbf{L}_2(G)$ which have first order generalized derivatives square-integrable over G. For a couple of functions $u, v \in W_2^1$ we define a scalar product by the formula

$$(u,v) = \int_G (\nabla u \cdot \nabla v + uv)dx \ . \tag{85}$$

If the norm $||u||_{W_2^1}$ of an element $u \in W_2^1$ is introduced as the nonnegative number

$$||u||_{W_2^1} = \left(\int_G (\nabla u \cdot \nabla u + u^2)dx\right)^{\frac{1}{2}} , \tag{86}$$

then the set W_2^1 is a separate Banach space.

The closure $\overset{\circ}{W}{}_2^1$ of the set of all compactly-supported in G functions with respect to the norm (86) is called a proper subspace of W_2^1.

The scalar product in $\overset{\circ}{W}{}_2^1$ can be defined by the formula

$$(u,v) = \int_G \nabla u \cdot \nabla v dx \ ,$$

instead of (85).

The following two assertions are valid.

(I) If we take the closure of the set of all compactly-supported functions with compact support in G with respect to the norm

$$||u|| = \left(\int_G \nabla u \cdot \nabla u dx\right)^{\frac{1}{2}} = (D(u))^{\frac{1}{2}} ,$$

then again we obtain the subspace $\overset{\circ}{W}{}_2^1$ of the space W_2^1.

(II) Every bounded set in $\overset{\circ}{W}{}_2^1$ is compact in $\mathbf{L}_2(G)$ (*Rellich's Theorem*).

The validity of the first assertion is obviously the consequence of the easily proved *Poincare-Steklov's inequality*

$$\int_G u^2 dx \le c^2 \int_G \nabla u \cdot \nabla u dx ,$$

where c is a constant which depends on the domain G only.

4.3.4. Now let the domain $G \subset E_n$ be bounded and its boundary $S = \partial G$ be an $(n-1)$-dimensional surface. We assume that G is in the specification domain of the uniformly ellipticity of linear second order equation

$$Lu \equiv \sum_{i,j=1}^{n} \frac{\partial}{\partial x_j} \left[A_{ij}(x) \frac{\partial u}{\partial x_i} \right] + \sum_{i=1}^{n} e_i \frac{\partial u}{\partial x_i} + cu = f(x) , \quad x \in G . \tag{87}$$

The classically formulated Dirichlet problem of equation (87) is to determine a regular solution $u(x)$ in G for equation (85) which is in the class $\mathbf{C}^{0,0}(G \cup S)$ by using the boundary condition

$$u(x) = g(x) , \quad x \in S , \tag{88}$$

where $g(x)$ is a given real continuous function on S.

The concept of a solution for Dirichlet's problem will be generalized in the case that the boundary condition (88) is homogeneous, i.e.,

$$u(x) = 0 , \quad x \in S . \tag{89}$$

The assumption that A_{ij}, e_i, c are bounded measurable functions and $f \in \mathbf{L}_2(G)$, what a *generalized solution* in the space W_2^1 for the problem (87)–(89) means is that a function $u(x) \in \overset{\circ}{W}{}_2^1$ for which the equality

$$\int_G \left(- \sum_{i,j=1}^{n} A_{ij} \frac{\partial u}{\partial x_i} \frac{\partial v}{\partial x_y} + \sum_{i=1}^{n} e_i \frac{\partial u}{\partial x_i} v + cuv - fv \right) dx = 0 \tag{90}$$

holds, for any function $v \in \overset{\circ}{W}{}_2^1$.

When a generalized solution $u(x) \in \mathbf{C}^{2,0}(G)$ and the functions A_{ij}, and surface S are smooth enough, we can rewrite the identity (90) in the form

$$\int_G (Lu - f)v dx = 0$$

after integrating by parts, from this and assuming the continuity of Lu and f we conclude that $u(x)$ is a classical solution of the problem (87)–(89).

Under the assumption that there is an adjoint operator with L:

$$L^*\omega \equiv \sum_{i,j=1}^{n} \frac{\partial}{\partial x_j}\left(A_{ij}\frac{\partial}{\partial x_i}\right) - \sum_{i=1}^{n}\frac{\partial}{\partial x_i}(e_i\omega) + cw \ ,$$

the function $w(x) \in \overset{\circ}{W}{}_2^1$, for which

$$\int_G \left(- \sum_{i,j=1}^{n} A_{ij}\frac{\partial w}{\partial x_i}\frac{\partial v}{\partial x_j} - \sum_{i=1}^{n}(e_i w)v + cwv \right) dx = 0 \tag{91}$$

holds for any function $v \in \overset{\circ}{W}{}_2^1$, is said to be a generalized solution in W_2^1 of the adjoint homogeneous equation

$$L^* w = 0 \ , \tag{92}$$

which satisfies the homogeneous boundary condition

$$w(x) = 0 \ , \quad x \in S \ . \tag{93}$$

The problem (92)–(93) is said to be *adjoint* to the problem (92)–(93).

Repeating the reasoning stated above on the basis of the identity (91), it can be verified that a generalized solution $w(x)$ from $\mathbf{C}^{2,0}(G)$ for the problem (92)–(93) is a classical solution.

In the theory of the problem (87)–(89), either in classical formulation or in the generalized one, the following assertions holds at the central position.

(a) The homogeneous Dirichlet problem (89) for the homogeneous equation

$$Lu = 0 \tag{94}$$

and its adjoint problem (92)–(93) have an identical finite number l of linearly independent solutions.

(b) The necessary and sufficient condition for the solvability of the problem (87)–(89) is that the functions $f(x)$ satisfies the conditions

$$\int_G f w_i dx = 0 \ , \quad i = 1, \dots , l \ ,$$

where $w_i, i = 1, \dots , l$, are all linearly independent solutions of the problem (92)–(93).

(c) The homogeneous problem (94)–(89) has only a trivial (identically equal to zero) solution in a domain G with sufficiently small measure.

An important theorem follows immediately from the assertions (a) and (b) as follows: the problem (87)–(89) for any $f(x)$ is solvable if and only if its corresponding homogeneous problem (89)–(94) has no nontrivial (different from zero) solutions.

On the basis of this theorem and the assertion (b), it follows that in a domain G with sufficiently small measure the problem (87)–(89) always has one and only one solution.

If there exists a principal elementary solution for equation (94), and S, f and the coefficients of equation (87) are sufficiently smooth, then, as we have already noticed in Chapter III that the potential method allows us to reduce the problem (87)–(88) to the equivalent *Fredholm's integral equation of the second kind.* Therefore the assertion (a), (b) and (c) are clearly valid in a classical formulation of the problem (87)–(88). For a generalized formulation of the problem (87)–(89) the validity of these assertions have also been proved by using functional methods. These methods can also be used to discuss other linear problems (Neumann's problem, problem with directional derivatives, etc.) which are generalized formulations for the uniformly elliptic equation (87).

We will say that linear boundary value problems for equation (87) have *Fredholm's property* if for them the assertions (a) and (b) are valid.

Remark 1. The homogeneous problems (89)–(94) and (92)–(93) are usually said to be the *associated* or *formally adjoint* problems.

The necessity mentioned above in the condition (b) for the solvability of the problem (87)–(89) in classical formulation also follows immediately from Green's formula (14) in Chapter III.

In fact, writing the above-mentioned formula using the notations $D = G, v = w$, and $d\tau_x = dx$, in the form

$$\int_G (wLu - uL^*w)dx = \int_S \left[a\left(w\frac{du}{dN} - u\frac{dw}{dN} \right) + buw \right] ds \ ,$$

and assuming that the problems (87)–(89) and (92)–(93) have solutions $u(x)$ and $w(x)$ respectively, we arrive at the equality

$$\int_G fwdx = 0 \ .$$

Remark 2. For the problem with a directional derivative discussed in §3.6 it is also possible to introduce the concept of associated (formally adjoint) problem, but such problem does not always have Fredholm's property. When in Chapter III the integer defined by the formula (116) $n \leq 0$, the nonhomogeneous problem with a directional derivative in some sufficiently smooth class is always solvable, and the corresponding homogeneous problem has $-2n+2$ real linearly independent solutions. When $n > 0$ the mentioned problem is solvable if the $2n-1$ real integral conditions (124) (according to the enumeration of the formulae in Chapter III) are fulfilled.

In the generalized Dirichlet problem (87)–(89), for the coefficients of equation (87) it was required that they are bounded and measurable and the function $f \in \mathbf{L}_2(G)$; but nothing is said about the smoothness of the boundary S of the domain G in which the solution is sought. Therefore if this problem has the solution $u(x)$ and it belongs to the space $\overset{0}{W}{}_2^1$, then it is difficult to judge in what sense it satisfies the boundary condition (89) without supplementary information about the smoothness of the surface S.

We denote by σ the part of S which is a bounded simply connected domain of the plane $x_n = 0$, and consider a cylinder G_0 of height h, with a generatrix parallel to the axis ox_n and basis σ belonging to G. Let $u(x)$ be a function with compact support in G.

From the obvious identity

$$u(x_1, \ldots, x_{n-1}, x_n) = \int_0^{x_n} \frac{\partial u(x_1, \ldots, x_{n-1}, t)}{\partial t} dt \ , \quad 0 \leq x_n \leq h \ ,$$

we obtain the estimate

$$\int_\sigma [u(x_1, \dots, x_{n-1}x_n)]^2 d\sigma \le x_n \int_G \left(\frac{\partial u}{\partial x_n}\right)^2 dx \ , \tag{95}$$

using Schwartz's inequality. This estimate is still valid for functions in $\overset{\circ}{W}{}_2^1$.

On account of (95) we conclude that the generalized solution $u(x)$ of the problem (87)–(89), if it exists, is such that the following equality holds,

$$\lim_{x_n \to 0} \int_\sigma [u(x_1, \dots, x_{n-1}, x_n]^2 d\sigma = 0 \ ,$$

when it is close to the section σ of the boundary S of the domain G.

It is always possible in principle to reduce a nonhomogeneous boundary Dirichlet's condition (88) to a homogeneous one if it is supposed that the boundary S of the domain G and the function $g(x)$ are smooth enough. For realizing this possibility it is sufficient, for example, to construct a function $u_0(x)$, harmonic in the domain G and satisfying the nonhomogeneous boundary condition (88), and then consider a new function $u_1(x) \equiv u(x) - u_0(x)$ which is a solution of the equation

$$Lu_1 = f - Lu_0 \ .$$

Evidently, the Dirichlet boundary condition for $u_1(x)$ is homogeneous.

It has been verified in Chapter III that for a well-posed Dirichlet's problem in the case of harmonic functions there is no necessity to require that its boundary value is a continuous function. In addition, as we have just seen, the generalized solution for the homogeneous Dirichlet problem (89) of equation (87) in the domain G must not have the value zero at every point $x \in S$ even if G has a sufficiently smooth boundary. By this reason, in the statement of Dirichlet's problem (87)–(88) the requirement on the continuity of $g(x)$ could be weakened, for example, to $g \in \mathbf{L}_2(s)$ instead, and define a generalized solution such that, finally, the formulated problem would be able to keep the Fredholm's property.

4.3.5. In both the classical and generalized Dirichlet's problem (87)–(88), as stated above, we assumed that the bounded domain G, in which the solution of this problem is to be sought, lies in the domain of the uniformly ellipticity of operator L. Now we explain by some examples to what extent this requirement is essential.

The equation in the Cartesian orthogonal coordinates x, y

$$y^2 \frac{\partial^2 u}{\partial x^2} + y \frac{\partial^2 u}{\partial y^2} - \frac{1}{2} \frac{\partial u}{\partial y} = 0 , \tag{96}$$

is elliptic in the upper half-plane $y > 0$. Let G be the finite domain in the half-plane $y > 0$, bounded by the arc σ of the semicircle $x^2 + \frac{4}{9}y^3 = 1$ in the Riemannian metric $ds^2 = ydy^2 + dx^2$, and the segment $A(-1,0)B(1,0)$ of the axis $y = 0$.

Because the corresponding characteristic form of equation (96) is given by the formula

$$Q(\lambda_1, \lambda_2) = y^2 \lambda_1^2 + y \lambda_2^2 ,$$

the uniform ellipticity condition

$$k_0(\lambda_1^2 + \lambda_2^2) \le Q(\lambda_1, \lambda_2) \le k_1(\lambda_1^2 + \lambda_2^2) , \quad 0 < k_0 = \text{const.} \le k_1 = \text{const.},$$

is not valid in G for this equation.

With the replacement of variables:

$$\xi = x , \quad \eta = \frac{2}{3} y^{\frac{3}{2}} ,$$

which is singular only when $y = 0$, the domain G becomes the half-disk G_1 : $\xi^2 + \eta^2 < 1, \eta > 0$, the acr σ becomes the semicircle $\sigma_1 : \xi^2 + \eta^2 = 1, \eta \ge 0$, the segment AB becomes $A_1(-1,0)B_1(1,0)$ of the axis of $\eta = 0$, the equation (96) becomes

$$v_{\xi\xi} + v_{\eta\eta} = 0 , \quad v(\xi, \eta) \equiv u \left[\xi, \left(\frac{3}{2} \eta \right)^{\frac{2}{3}} \right] , \tag{97}$$

and Dirichlet's boundary condition

$$u(x, y) = g(x, y) , \quad (x, y) \in \partial G \tag{98}$$

for the equation (96) becomes the boundary condition

$$v(\xi, \eta) = g \left[\xi, \left(\frac{3}{2} \eta \right)^{\frac{2}{3}} \right] , \quad (\xi, \eta) \in \partial G_1 \tag{99}$$

for equation (97).

Because Dirichlet's problem (97)–(99) is well-posed (its solution is constructed by quadrature), so is the problem (96)–(98).

In the finite domain G in the upper half-plane $y > 0$ bounded by a simple Jordan arc σ with endpoints $A(a,0)$ and $B(b,0), a < b$ and a segment AB of the axis $y = 0$, the equation

$$Lu = \frac{\partial^2 u}{\partial x^2} + y\frac{\partial^2 u}{\partial y^2} + \frac{\partial u}{\partial y} - u = 0 \tag{100}$$

is elliptic, but still does not satisfy the condition of uniform ellipticity.

The problem to find a solution $u(x,y)$ of equation (100), which is regular in G, being in the class $\mathbf{C}^{0,0}(G \cup \sigma)$, is bounded when $y \to 0$ and satisfies the Dirichlet boundary condition on the section σ of the boundary ∂G of the domain G,

$$u(x,y) = g(x,y) \; , \quad (x,y) \in \sigma \; , \tag{101}$$

can not have more than one solution.

To justify this assertion, it is enough to show that the started problem with the homeogeneous boundary condition

$$u(x,y) = 0 \; , \quad (x,y) \in \sigma \; , \tag{102}$$

has no nontrivial solutions.

Let $u(x,y)$ be a solution of the problem (100)–(102) and consider the function $w = -\log y + k$, where by $\log y$ we mean the branch of this function which is real for $y > 0$, and the positive constant k is chosen such that $w > 0$ for $(x,y) \in G \cup \partial G$.

It is obvious that

$$Lw = -w < 0 \tag{103}$$

at every point of the domain G. The function $v(x,y) = \varepsilon w \pm u$, where ε is an arbitrary positive constant, is positive on ∂G. It cannot take a negative minimum in the domain G. In fact, suppose that the function $v(x,y)$ reaches a negative relative minimum at a point $(x,y) \in G$, then we should have $Lv = v_{xx} + yv_{yy} - v > 0$. But by (103) this is impossible:

$$Lv = \varepsilon Lw = -\varepsilon w < 0 \; .$$

Therefore, there is the estimate $\varepsilon w \pm u > 0$ at every fixed point $(x, y) \in G$, i.e. $|u(x, y)| < \varepsilon w$. From this we obtain $u(x, y) = 0$ on account of the arbitrariness of the ε, and the uniqueness of the solution for the problem (100)–(101) is therefore proved. It can also be proved that this problem has a solution for an arbitrary continuous function $g(x, y)$.

The elliptic equation

$$\frac{\partial^2 u}{\partial x^2} + y\frac{\partial^2 u}{\partial y^2} + \frac{1}{2}\frac{\partial u}{\partial y} = 0 \tag{104}$$

is not uniformally elliptic in any domain G which lies in the upper half-plane $y < 0$, adjoining the axis $y = 0$.

For G we take the finite domain bounded by the parabola $x^2 + 4y = 1, y > 0$, and the segment $A(-1, 0)B(1, 0)$ of the axis $y = 0$.

By the change of variables.

$$\xi = x\ , \quad \eta = 2y^{\frac{1}{2}}\ , \quad v(\xi, \eta) \equiv u\left[\xi, \frac{1}{4}\eta^2\right]\ ,$$

equation (104) becomes Laplace's equation (97), and the domain G the half-disk $G_1 : \xi^2 + \eta^2 < 1, \eta > 0$, in the plane of the variables ξ and η.

If we require that the derivative $\frac{\partial u}{\partial y}$ to exist in the domain G and is square integrable, then we have

$$\lim_{\eta \to 0} \frac{\partial v}{\partial \eta} = \lim_{y \to 0} y^{\frac{1}{2}} \frac{\partial u}{\partial y} = 0\ ,$$

by virtue of the equality

$$\frac{\partial u}{\partial y} = y^{\frac{1}{2}} \frac{\partial u}{\partial \eta}\ .$$

Under these assmuptions, if $u(x, y) \in \mathbf{C}^{0,0}(G \cup \partial G)$ we conclude that the function $v(\xi, \eta)$ which is harmonic in G_1 is harmonically extended to the half-disk $\overline{G}_1 : \xi^2 + \eta^2 < 1, \eta < 0$, through the segment $A_1(-1, 0)B_1(1, 0)$, and hence the boundary condition (101) for $u(x, y)$ becomes Dirichlet's boundary condition for the function $v(\xi, \eta)$,

$$\begin{aligned} v(\xi, \eta) &= g_1(\xi, \eta)\ , \quad (\xi, \eta) \in \sigma_1\ , \\ v(\xi, \eta) &= g_1(\xi, -\eta)\ , \quad (\xi, \eta) \in \overline{\sigma}_1\ , \end{aligned} \tag{105}$$

where

$$g_1(\xi,\eta) = g\left[\xi, \left(\frac{1}{2}\eta\right)^2\right] .$$

The arc σ_1 which lies in the half-plane $\eta \geq 0$ is a part of ∂G_1, and the arc $\bar{\sigma}_1$ is symmetric to it with respect to the segment A_1B_1.

Since Dirichlet's problem (97)–(105) is uniquely solvable, the problem (101)–(104) is also uniquely solvable under the assumptions adopted above.

4.3.6. As we have already noticed, in the case of the linear scalar partial differential equation of second order, (87), the uniform ellipticity guarantees the Fredholm property of Dirichlet's problem (88) in a bounded domain. But this may not be true in the case of a second order elliptic system.

In fact, let us consider the system of two equations

$$\begin{aligned} &\frac{\partial^2 u_1}{\partial x^2} - \frac{\partial^2 u_1}{\partial y^2} - 2\frac{\partial^2 u^2}{\partial x \partial y} = 0 , \\ &2\frac{\partial^2 u_1}{\partial x \partial y} + \frac{\partial^2 u_2}{\partial x^2} - \frac{\partial^2 u_2}{\partial y^2} = 0 , \end{aligned} \tag{106}$$

written in matrix form

$$Au_{xx} + 2Bu_{xy} + cu_{yy} = 0 , \tag{107}$$

where

$$A = \begin{pmatrix} 1 & 0 \\ 0 & 1 \end{pmatrix} , \quad B = \begin{pmatrix} 0 & -1 \\ 1 & 0 \end{pmatrix} , \quad c = \begin{pmatrix} -1 & 0 \\ 0 & -1 \end{pmatrix} ,$$

and $u = (u_1, u_2)$.

This system is uniformly elliptic because the corresponding characteristic form of the system (107) is

$$k(\lambda_1, \lambda_2) = \det(A\lambda_1^2 + 2B\lambda_1\lambda_2 + c\lambda_2^2) = (\lambda_1^1 + \lambda_2^2)^2 ,$$

and hence

$$\begin{gathered} k_0(\lambda_1^2 + \lambda_2^2)^2 \leq k(\lambda_1, \lambda_2) \leq k_1(\lambda_1^2 + \lambda_2^2)^2 \\ 0 \leq k_0 \, \text{const.} \leq 1 , \quad 0 \leq k_1 = \text{const.} \geq 1 . \end{gathered}$$

It is easy to see that in the disk $G : |z - z_0|^2 < \varepsilon^2$, in spite of the arbitrary location of its centre z_0 on the plane of complex variable $z = x + iy$ and the arbitrary finite radius ε, the homogeneous Dirichlet problem of the system (106)

$$\begin{aligned} u_1(x,y) &= 0 , \\ u_2(x,y) &= 0 , \\ (x-x_0)^2 + (y-y_0)^2 &= \varepsilon^2 \end{aligned} \tag{108}$$

has an infinite set of linearly independent solutions

$$w_0(z) = u_1(x,y) + iu_2(x,y) = [(z - z_0)(\overline{z} - \overline{z}_0) - \varepsilon^2]\varphi(z) , \tag{109}$$

where $\varphi(z)$ is an arbitrary function of complex variable z which is analytic in G.

In fact, by the notations

$$\overline{z} = x - iy , \quad w(z) = u_1(x,y) + iu_2(x,y) , \quad \frac{\partial}{\partial \overline{z}} = \frac{1}{2}\left(\frac{\partial}{\partial x} + i\frac{\partial}{\partial y}\right) ,$$

the system (106) has the form

$$\frac{\partial^2 w}{\partial \overline{z}^2} = 0 ,$$

from which it follows that

$$w(z) = \overline{z}\phi(z) + \psi(z) , \tag{110}$$

where $\phi(z)$ and $\psi(z)$ are arbitrary analytic functions of complex variable z in G.

On the basis of (110) we conclude that the function $w_0(z)$ expressed by the formula (109) is a regular solution of the system (106). The functions $u_1(x,y) = \mathrm{Re} w_0(z)$ and $u_2(x,y) = \mathrm{Im} w_0(z)$ obviously satisfy the homogeneous boundary conditions (108).

Therefore the Dirichlet's problem in circle G for the system (106) is not Fredholmian.

At present there have been found the conditions strengthening the ellipticity of a linear system of partial differential equations of second order, and their fulfillment guarantees the Fredholm property of Dirichlet's problem.

4.3.7. The concepts of generalized solutions are introduced not only for elliptic type equations.

Let us consider the hyperbolic type equation

$$\frac{\partial^2 u}{\partial t^2} - Lu = f(x,t) , \tag{111}$$

where the operator

$$Lu \equiv \sum_{i,j=1}^{n} \frac{\partial}{\partial x_j}\left[A_{ij}(x,t)\frac{\partial}{\partial x_i}\right] + \sum_{i=1}^{n} e_i(x,t)\frac{\partial u}{\partial x_i} + C(x,t)u$$

is uniformly elliptic and positive in its own specification domain Ω of the space E_{n+1} of points (x,t). Let G be a bounded domain in the space E_n of points $(x,0)$, and $Q : \{G \times (0 < t < T)\}$ a subdomain of Ω with bottom G and lateral surface $S : \{\partial G \times (0 \leq t \leq T)\}$, with $T =$ constant > 0.

The requirement of determining a solution $u(x,t)$ of equation (111) which is regular in Q, continuous on $Q \cup \partial Q$, and satisfies the conditions

$$\begin{aligned} &u(x,0) = \tau(x) , \quad \left.\frac{\partial u(x,t)}{\partial t}\right|_{t=0} = \nu(x) , \quad x \in G , \\ &u_s = 0 , \quad x \in S \end{aligned} \tag{112}$$

is called the *first basic mixed problem* for equation (111).

The problem discussed before, (5)–(6)–(7), is a special case of the problem (111)–(112).

The formulation of this problem can be also generalized by making use of the concept of generalized derivatives.

By W^1_{20} we denote the closure with respect to the norm of $W^1_2(Q)$ of functions which are smooth in Q and vanish near S.

A function $u(x,t) \in W^1_{20}$ is called a *generalized solution* of the problem (111)–(112) in the space $W^1_{20}(Q)$ if it becomes $\tau(x) \in W^1_2(G)$ when $t = 0$ and satisfies the identity

$$\begin{aligned} &\int_Q \left(-\frac{\partial u}{\partial t}\frac{\partial v}{\partial t} + \sum_{i,j=1}^{n} A_{ij}\frac{\partial u}{\partial x_i}\frac{\partial v}{\partial x_j} - \sum_{i=1}^{n} e_i \frac{\partial u}{\partial x_i} v - cuv \right) dxdt \\ &= \int_G \nu(x)v(x,0)dx + \int_Q fvdxdt \end{aligned} \tag{113}$$

for every function $v(x,t) \in W^1_{20}(Q)$ which is equal to zero at $t = T$.

For the identity (113) to make sense it is enough to require A_{ij}, e_i, c to be bounded and measurable and $f \in \mathbf{L}_2(Q), \nu \in \mathbf{L}_2(G)$.

It $\tau(x) \equiv 0$ the identity obviously holds for a classical solution $u(x,t)$ of the problem (111)–(112) under some additional assumptions on the smoothness of coefficients of equation (111), the function $\nu(x)$ and the boundary of the domain Q.

Before the definition of the generalized solution for the problem (111)–(112) that we have just stated was formulated there had been historically another definition.

Let $\{f_k(x,t)\}, k = 1, 2, \dots$, be a bounded sequence belonging to the function space $L_2(G)$, the limit function of which is the function $f(x,t)$ for each fixed $t, 0 \leq t \leq T$.

If for every k the problem (111)–(112) has a unique classical solution $u_k(x,t) \in \mathbf{L}_2(Q)$ when $f = f_k$, and the function $u(x,t)$ is the limit of the sequence $\{u_k(x,t)\}$ in $\mathbf{L}_2(G)$ for each fixed $t, 0 \leq t \leq T$, then this function is called a generalized solution of the problem (111)–(112).

4.3.8. Suppose that we now have a parabolic type equation

$$u_t - Lu = f(x,t) , \tag{114}$$

where the differential operator Lu and the right-hand side f satisfy the same conditions as in equation (111). We denote by G, as above, a bounded domain in the space E_n, and by $Q : \{G \times (0 < t < T)\}$, a subdomain of the specification domain Ω of equation (114).

The classical first boundary value problem for equation (114) is to determine a regular solution $u(x,t)$ in Q for equation (114) such that

$$u(x,t) \in \mathbf{C}^{0,0}(Q \cup \partial Q)$$

and it satisfies the conditions

$$u(x,0) = \varphi(x) , \quad x \in G , \tag{115}$$

$$u(x_0,t) = 0 , \quad (x_0,t) \in S , \tag{116}$$

where $S : \{\partial G \times (0 \leq t \leq T)\}$.

A generalized solution of the problem (114)–(115)–(116) can be defined in a different way.

If the scalar product (u,v) of two functions $u(x,t)$ and $v(x,t) \in \mathbf{L}_2(Q)$ which have generalized first order derivatives the spatial variables from $\mathbf{L}_2(Q)$ is defined as the integral form

$$(u,v) = \int_Q \left(\sum_{i=1}^{n} \frac{\partial u}{\partial x_i} \frac{\partial v}{\partial x_i} + uv \right) dx dt$$

and the norm of $u(x,t)$ is given by the formula

$$\|u\| = (u,u)^{\frac{1}{2}} ,$$

then we obtain a Hilbert space $\mathbf{W}_2^{10}$. The closure $\overset{\circ}{W}{}_2^{10}$, of all functions which are smooth on $Q \cup \partial Q$ and vanish near S, in the norm of the space W_2^{10} is a subspace of the space W_2^{10}.

Under the assumption $\varphi \in \mathbf{L}_2(G)$, we say that a function $u(x,t) \in \overset{\circ}{W}{}_2^{10}$ is a generalized solution of the problem (114)–(115)–(116) in the space W_2^{10} if the identity

$$\int_Q \left(-u\frac{\partial v}{\partial t} + \sum_{i,j=1}^{n} A_{ij} \frac{\partial u}{\partial x_i} \frac{\partial v}{\partial x_j} - \sum_{i=1}^{n} e_i \frac{\partial u}{\partial x_i} v - cuv \right) dx dt$$
$$= \int_G \varphi(x) v(x,0) dx + \int_Q f v dx dt \tag{117}$$

holds for every function $v(x,t) \in \overset{\circ}{W}{}_2^{10}$ which vanishes at $t = T$.

The classical solution $u(x,t)$ of the problem (114)–(115)–(116), of course, satisfies the identity (117).

If the functions $f(x,t)$ and $\varphi(x)$ are the limits of the sequences of functions $\{f_k(x,t)\}$ and $\{\varphi_k(x)\}, k = 1, 2, \dots$, which are uniformly convergent in $Q \cup \partial Q$ and $G \cup \partial G$ respectively and sufficiently smooth. Furthermore, problem (114)–(115)–(116) has a unique classical solution $u_k(x,t)$ when f and φ are replaced by $f_k(x,t)$ and $\varphi_k(x)$ respectively for every k. Then, provided the sequence $\{u_k(x,t)\}$ is uniformly convergent in $Q \cup \partial Q$, their limit $u(x,t)$ is also called the generalized solution of the problem (114)–(115)–(116).

Thus, the introduction of generalized solutions is justified, above all, in its application (for example, in the linear theory of heat conductivity) to functions $f(x,t)$ and $\varphi(x)$ which are usually given with to a specific accuracy.

§4.4 A Brief Survey of Methods for Proving Existence of Generalized Solutions

4.4.1. Below we only investigate some examples of nonhomogeneous partial differential equations with homogeneous boundaries and initial conditions. It can always be achieved by using a replacement for the unknown function, in principle, to make the boundary and initial conditions homogeneous, which corresponds to an alteration of the right-hand side of the equation. To demonstrate the correctness of this assertion we strengthen the requirement on smoothness of the coefficients of the equation, the given functions and their supports.

In accordance with the above definition, a function $u(x)$ given in a bounded domain $G \in E_n$ and taken from the Hilbert space $\overset{\circ}{W}{}_2^1$ is called the generalized solution of the homogeneous Dirichlet problem

$$u(x) = 0\ , \quad x \in \partial G\ , \tag{118}$$

for Poisson's equation

$$\Delta u = -f(x)\ , \quad f \in \mathbf{L}_2(G)\ , \tag{119}$$

if it satisfies the identity

$$\int_G (\nabla u \nabla v - fv)dx = 0 \tag{120}$$

for every function $v(x) \in \overset{\circ}{W}{}_2^1$.

As we have noticed, the scalar product in $\overset{\circ}{W}{}_2^1$ can be written in the form

$$(u, v) = \int_G \nabla u \cdot \nabla v dx$$

and in the meantime this space is obtained by taking the closure of all the functions with a compact support in G with respect to the norm

$$||u|| = \left(\int_G (\nabla u)^2 dx\right)^{\frac{1}{2}}\ .$$

The expression

$$F(v) = \int_G fv dx \tag{121}$$

is a bounded linear functional defined in the Hilbert space $\overset{\circ}{W}{}_2^1$. From the well-known proposition in functional analysis (the *Riesz representation theorem* of linear functional) there exists a unique function $u(x) \in \overset{\circ}{W}{}_2^1$ such that the functional (121) can be represented in the form

$$F(v) = (u, v) = \int_G \nabla u \cdot \nabla v dx \tag{122}$$

for every function $v \in \overset{\circ}{W}{}_2^1$. The existence and uniqueness of a generalized solution for the problem (118)–(119) follows from the "identityness" of the identities (120) and (122).

We are not going to turn our attention to the very interesting and difficult problem of the coincidence of the generalized solutions with the classical ones of the problem (118)–(119) here, but only to point out that this coincidence is strictly true only if $f(x)$ and ∂G are sufficiently smooth.

4.4.2. Let $Q\{G \times (0 < t < T)\}$ be the domain in the space E_{n+1} that we have just introduced in the preceding paragraph and G a bounded domain of the space E_n. By $\{f_k(x,t)\}$ we denote a sequence of bounded functions given in Q and the function $f(x,t)$ which it converges its limit to denote function in $\mathbf{L}_2(G)$ to uniformly with respect to $t, 0 \leq t \leq T$. We suppose that when $f(x,t) = f_k(x,t), k = 1, 2, \ldots$ of the basic homogeneous mixed problem in the domain Q

$$u(x,0) = 0\,, \quad u_t(x,0) = 0\,, \quad x \in G\,, \quad u|_s = 0\,, \quad s = \{\partial G \times (0 \leq t \leq T)\}\,, \tag{123}$$

for the homogeneous wave equation

$$\frac{\partial^2 u}{\partial t^2} - \Delta u = f(x,t) \tag{124}$$

there always exists a unique classical solution $u_k(x,t)$ for every k. It is assumed that the boundary ∂Q of the domain Q and the function $u_k(x,t)$ in $Q \cup \partial Q$ are smooth enough so that the integral changes that will be used below are valid.

According to the definition stated above, the function $u(x,t)$ which is the limit of the sequence $\{u_k(x,t)\}$ in $\mathbf{L}_2(G)$ for every $t, 0 \leq t \leq T$, is called a generalized solution of the problem (123)–(124). We now prove the existence of such a generalized solution for the problem (123)–(124).

For this we introduce for consideration the *energy integral*

$$E_k^2(t) = \frac{1}{2}\int_G \left[\left(\frac{\partial u_k}{\partial t}\right)^2 + (\nabla u_k)^2\right] dx \ . \tag{125}$$

It is easy to show that

$$E_k^2(t) = \int_0^t dt_1 \int_G \frac{\partial u_k(x,t_1)}{\partial t_1} f_k(x,t_1) dx \ . \tag{126}$$

In fact, we have for $u_k(x,t)$ the identity

$$u_k(x,0) = u_{kt}(x,0) = 0 \ , \quad x \in G \ , \quad u_k|_S = 0 \ , \tag{123_k}$$

$$\frac{\partial^2 u_k(x,t_1)}{\partial t_1^2} - \Delta u_k(x,t_1) = f_k(x,t_1) \ , \quad (x,t_1) \in Q \ . \tag{124_k}$$

Multiplying both sides of (124_k) by $\frac{\partial u_k(x,t_1)}{\partial t_1}$ and integrating them over the domain $Q_t : \{G \times (0 < t_1 < t)\}$ we obtain

$$\begin{aligned} &\int_0^t dt_1 \int_G \left[\frac{1}{2}\frac{\partial}{\partial t_1}\left(\frac{\partial u_k}{\partial t_1}\right)^2 + \nabla u_k \nabla \frac{\partial u_k}{\partial t_1} - \sum_{i=1}^n \frac{\partial}{\partial x_i}\left(\frac{\partial u_k}{\partial t_1}\frac{\partial u_k}{\partial x_i}\right)\right] dx \\ &= \int_0^t dt_1 \int_G f_k(x,t_1) \frac{\partial u_k(x,t_1)}{\partial t_1} dx \ , \end{aligned} \tag{127}$$

where the gradient operator ∇ is taken with respect to the spatial variables $x_1, \ldots, x_n$.

By exchanging the order of integrations and applying the formula (G-O) the equality (127) has the form

$$\begin{aligned} &\frac{1}{2}\int_G \left[\left(\frac{\partial u_k}{\partial t_1}\right)^2 + (\nabla u_k)^2\right]_{t_1=0}^{t_1=t} dx - \int_0^t dt_1 \int_{\partial G} \frac{\partial u_k}{\partial t_1}\frac{\partial u_k}{\partial N_x} ds_x \\ &= \int_0^t dt_1 \int_G f_k(x,t_1) \frac{\partial u_k(x,t_1)}{\partial t_1} dx \ , \end{aligned} \tag{128}$$

where N_x is the unit outward normal vector to ∂G at the point x. when $t_1 = 0$, by the first two of the conditions (123_k) we have

$$\frac{\partial u_k}{\partial t_1} = 0 \ , \quad \nabla u_k = 0 \ , \quad x \in G \tag{129}$$

and from the last one of these same conditions we get

$$\int_0^t dt_1 \int_{\partial G} \frac{\partial u_k}{\partial t_1}\frac{\partial u_k}{\partial N_x} ds_x = \int_S \frac{\partial u_k}{\partial t_1}\frac{\partial u_k}{\partial N_x} ds = 0 \ . \tag{130}$$

Taking into consideration (129) and (130) the equality (128) becomes the equality (126).

By differentiating with respect to t we obtain from (126)

$$2E_k(t)\frac{d}{dt}E_k(t) = \int_G \frac{\partial u_k(x,t)}{\partial t} f_k(x,t)dx \ ,$$

and from this we arrive at the estimate

$$2E_k(t)\frac{d}{dt}E_k(t) \le \left\| \frac{\partial u_k}{\partial t} \right\| \|f_k\| \ , \tag{131}$$

on account of Schwarz's inequality, where

$$\left\| \frac{\partial u_k}{\partial t} \right\|^2 = \int_G \left(\frac{\partial u_k}{\partial t} \right)^2 dx \ , \quad \|f_k\|^2 = \int_G (f_k)^2 dx \ .$$

The estimate

$$\left\| \frac{\partial u_k}{\partial t} \right\| \le \sqrt{2} E_k(t) \tag{132}$$

is obtained immediately from (125). On the basis of (132) and (131),

$$\frac{dE_k(t)}{dt} \le \frac{1}{\sqrt{2}} \|f_k\| \ .$$

By integration, the estimate

$$E_k(t) \le \frac{1}{\sqrt{2}} \int_0^t \|f_k(t_1)\| dt_1 \tag{133}$$

follows, since $E_k(0) = 0$ in view of (126).

Considering (133) the estimate (132) has the form

$$\left\| \frac{\partial u_k}{\partial t} \right\| \le \int_0^t \|f_k(t_1)\| dt_1 \ . \tag{134}$$

As

$$||u_k(t)||^2 = \int_G u_k^2 dx \tag{135}$$

for every t in $0 \le t \le T$,

$$\begin{aligned} 2||u_k|| \frac{d}{dt} ||u_k|| &= 2 \int_G u_k \frac{\partial u_k}{\partial t} dx \\ &\le 2||u_k|| \left\| \frac{\partial u_k}{\partial t} \right\| . \end{aligned} \tag{136}$$

Making use of the estimate (134) we obtain from (136)

$$\frac{d}{dt} ||u_k|| \le \int_0^t ||f_k(t_1)|| dt ,$$

or, taking into consideration the fact that $||u_k(0)|| = 0$ due to the first of the conditions (123_k) and (135),

$$||u_k(t)|| \le \int_0^t dt_1 \int_0^{t_1} ||f_k(t_2)|| dt_2 = \int_0^t (t - t_1) ||f_k(t_1)|| dt_1 . \tag{137}$$

Particularly, the boundedness of the sequence $\{u_k(x, t)\}$ in $\mathbf{L}_2(G)$ follows from that of $\{f_k(x, t)\}$ for every $t, 0 \le t \le T$.

Since the sequence $\{f_k(x, t)\}$ converges in $\mathbf{L}_2(G)$ uniformly with respect to $t, 0 \le t \le T$, for arbitrary $\varepsilon < 0$ there exists a number $N > 0$ such that

$$||f_{N+p} - f_N|| < \varepsilon \tag{138}$$

for all $p > 0$.

On the basis of (138), we obtain from (137) that for each $t, 0 \le t \le T$,

$$||u_{N+p}(x, t) - u_N(x, t)|| \le \frac{\varepsilon}{2} t^2 . \tag{139}$$

The estimate (139) means that for each $t, 0 \le t \le T$, the sequence $\{u_k(x, t)\}$ is fundamental in $\mathbf{L}_2(G)$. Therefore, according to the well-known *Fisher-Ritz theorem* in functional analysis, this sequence converges in $\mathbf{L}_2(G)$ and its limit in $\mathbf{L}_2(G)$ is a completely definite function $u(x, t)$ for each $t, 0 \le t \le$

T. By the definition given above, the function $u(x,t)$ is a generalized solution of the problem (123)–(126). The uniqueness of the solution is obvious.

4.4.3. Now suppose that

$$\{f_k(x,t)\}\ , \quad k=1,2,\dots\ , \tag{140}$$

is a uniformly convergent sequence of functions which are continuous on $Q \cup \partial Q$ and whose limit we denote by $f(x,t)$.

It is supposed that for each k there exists a unique classical solution $u_k(x,t)$ which is continuous on $Q \cup \partial Q$ for the first boundary value problem,

$$u_k(x,0)=0\ , \quad x \in G\ ; \quad u_k|_s = 0\ , \tag{141}$$

of the homogeneous heat-conduction equation

$$\frac{\partial u}{\partial t} - \Delta u_k = f_k(x,t)\ , \quad (x,t) \in Q\ . \tag{142}$$

We show that the sequence

$$\{u_k(x,t)\}\ , \quad k=1,2,\dots\ , \tag{143}$$

uniformly converges in $Q \cup \partial Q$ and, hence, its limit $u(x,t)$ is a generalized solution of the first boundary problem

$$\frac{\partial u}{\partial t} - \Delta u = f(x,t)\ , \quad (x,t) \in Q\ ,$$
$$u(x,0)=0\ , \quad x \in G\ , \quad u|_s = 0\ ,$$

stated at the end of the preceding paragraph.

First of all we notice that for the solution $u_k(x,t)$ of the problem (141)–(142) the estimate

$$|u_k(x,t)| \le TM_k \tag{144}$$

is valid at every point $(x,t) \in Q \cup \partial Q$, provided that

$$\max_{(x,t)\in Q\cup\partial Q} |f_k(x,t)| < M_k\ . \tag{145}$$

To prove this assertion let us at first suppose that conversely there exists a point (x_0, t_0) in the domain Q or on its upper base where

$$u_k(x_0, t_0) > TM_k \ . \tag{146}$$

This hypothesis will lead to a contradiction. In fact, due to (146) the function

$$v(x,t) = u_k(x,t) + M_k(T-t)$$

must arrive at its own maximum in $Q \cup \partial Q$ at some point in Q or on its upper base. But this is impossible, from (145),

$$\Delta v - v_t = -f_k(x,t) + M_k > 0$$

at the point (x,t). The assumption that $u_k(x,t) < -TM_k$ will lead to contradiction also.

Since the sequence (140) is fundamental, for arbitrary $\varepsilon > 0$ there exists a number $N > 0$ such that for all $(x,t) \in Q \cup \partial Q$ and arbitrary positive p,

$$|f_{N+p} - f_N| < \varepsilon \ . \tag{147}$$

In view of the fact that

$$\left(\frac{\partial}{\partial t} - \Delta\right)(u_{N+p} - u_N) = f_{N+p} - f_N$$

and

$$U_{N+p}(x,0) - u_N(x,0) = 0 \ , \quad x \in G \ , \quad [u_{N+P} - u_N]_S = 0 \ ,$$

from the assertion proved above and the estimate (147) it follows that for all $(x,t) \in Q \cup \partial Q$ and arbitrary $p < 0$ the estimate

$$|u_{N+p}(x,t) - u_N(x,t)| \leq T\varepsilon$$

holds, i.e., the sequence (143) is fundamental in the space $\mathbf{C}^{0,0}(Q \cup \partial Q)$. Therefore, this sequence converges uniformly, and its limit $u(x,t)$ is a continuous function, being a generalized solution of the problem (141)–(142) by definition.

The conclusion deduced here does not make it possible to judge the differentiability of the function $u(x,t)$.

When investigating the problem of smoothness of the generalized solution for hyperbolic and parabolic equations certain difficulties are encountered, which could not always be overcome successfully.

Above we have given essentially two definitions of the generalized solution for partial differential equations. The first was presented after the introduction of generalized derivatives of classical functions, and the generalized solution was an element of the perfectly definite Banach spaces $\overset{\circ}{W}{}_2^1, W_{20}^1$ or $\overset{\circ}{W}{}_2^{10}$. The idea of introducing the second definition comes from the interpretation of the generalized solution as the limit of classical solutions in the norm of the appropriately selected metric space. This idea has been further developed to generalization of concepts concerning not only the derivatives but also the function itself. On the basis of this a whole theory has been developed as the so-called theory of generalized functions and distributions in the last fifty years.

Chapter V. Some Classes of Non-linear Partial Differential Equations

§5.1 General Observations

5.1.1. In all previous chapters except chapter I (the introduction) we only mentioned linear partial differential equations and their linear (Dirichlet's, Neuman's, Cauchy's characteristic and mixed) problems. The basic hydromechanical equations (the Navier-Stockes's equation (24) and the continuity equation (25)) and also equations in the general theory of relativity (Einstein's equations (32)) written out in Chapter I are nonlinear partial differential equations. They reduce to linear equations under very strong, but not always justifiable, conditions. In the same chapter some assumptions have been stated under which linear equations could be derived to describe propagation of sound, transmission of heat, equilibrium (static) stress states of an elastic medium, and electrostatic fields. We also sated the linear (Cauchy's, Dirichlet's, Neuman's and various mixed) problems corresponding to these equations. Occasionally the situation forced us to make observations on nonlinear problems also. For example, it was proved in §4 of Chapter II that the planar stationary irrotational motion of a non-viscous compressible medium is described by a system of nonlinear first order partial differential equations (38) in terms of the potential of velocity $\varphi(x_1, x_2)$ and the stream function $\psi(x_1, x_2)$.

For the system (38) problems are presented in two types on the physical plane: with a fixed boundary and with a free boundary. Let D_0 be an infinite domain of the plane of variables x_1 and x_2, which is either (a) the exterior of a closed piecewise smooth Jordan boundary S_0 (which is the case when observing flows around airfoils), or (b) a part of this plane located on a side of a curve S_0 with two endpoints at infinity, or (c) a band with two boundary components S_1 and S_2.

The simplest problem with a fixed boundary is to determine a solution φ, ψ of the system (38) when it is known that there is no flow passing through S_0, namely,

$$\frac{\partial\varphi(x_1, x_2)}{\partial\nu} = 0 \ , \quad (x_1, x_2) \in S_0 \ , \tag{1}$$

where ν is the normal to S_0 at the point (x_1, x_2). By virtue of (38) it follows from (1) that

$$\psi(x_1, x_1) = \text{const.} \ , \quad (x_1, x_2) \in S_0 \ . \tag{2}$$

In addition to the conditions (1) and (2) it is assumed that at infinity the absolute value of the speed is given

$$|q| = \frac{\rho_0}{\rho}|\nabla\psi|$$

and, besides, the equality

$$\int_C d\varphi = 0$$

holds for the case (a), where the integration is taken along an arbitrary closed piecewise smooth path C which is in D_0 and encloses S_0.

The problems (1) and (2) are obviously linear.

It has been found that if only the definite section S_1 of the boundary ∂D_0, which is the moving part of the compressible medium D_0 is known, then we should find the remaining part, S_2, where $S_1 \cup S_2 = \partial D_0$, together with the solution of the system (38). The problem of searching for the section S_2 of the boundary of the domain D_0 and the solution φ, ψ of the system (38) in the domain D_0 according to the boundary condition

$$\psi(x_1, x_2) = \text{const.}, \quad (x_1, x_2) \in S_0 ,$$
$$q = \text{const.}, \quad (x_1, x_2) \in S_2 ,$$

is called a *problem with free boundary.*

Although the nonlinear system (38) can be reduced to a linear system (45) by replacement of the independent variables using the equality (41) (the transformation of hodograph), we need to know the domain D_0 and its boundary S_0 on the hodograph plane in advance to write the conversion formula $x_1 = x_1(q, \vartheta), x_2 = x_2(q, \vartheta)$. Therefore, the linear boundary conditions (1) and (2) of the functions φ and ψ generate the nonlinear boundary conditions for the solutions $u_1(\vartheta, v)$ and $u_2(\vartheta, v)$ of the system (45) on the image of ∂D_0, and hence we have to deal with a nonlinear boundary value problem of the system (45), or the Chaplygin's equation

$$K(v)u_{\vartheta\vartheta} + u_{vv} = 0 , \quad u = u_2 \tag{3}$$

in a domain of the plane of variables ϑ and v, which is the image of the domain D_0 under the hodograph transformation.

It is far from being true that every solution $u(\vartheta, v)$ of the equation (3) always corresponds to a real motion on the physical plane. In this respect it will be very difficult when investigating a transonic motion because the domain D_0 which is occupied by the media should correspond to a domain D on the plane of variables ϑ and v, the intersections of which with either the upper half-plane $v > 0$ or the lower half-plane $v < 0$ are not all empty (ϑ and v are understood to be Cartesian orthogonal coordinates). In the domain D the equation (3) belongs to the class of mixed type equations. If $K(v) = v^m \mathrm{sgn} v$, where m is a nonnegative number, then the equation (3) is usually called a *model equation of mixed type.*

Today the model equations of mixed type have been investigated well enough. For simplicity we shall only mention a class of boundary problems of these equations.

5.1.2. If $v < 0$ the model equation (3) is hyperbolic and it has two families of characteristic curves defined by the equations

$$d\vartheta \pm \sqrt{-k(v)}dv = 0 \ .$$

Let D be a finite simply-connected domain of the plane of variables ϑ and v which is bounded by a simple Jordan arc σ with endpoints $A(0,0)$ and $B(1,0)$, in the upper half-plane $v > 0$, and by the sections AC and BC of the different characteristic families of the equation (3), which are emanating from a fixed point $C(\vartheta_0, v_0)$ in the lower half-plane $v < 0$.

The problem to determine, in a domain D a regular solution $u(\vartheta, v)$ of the equation (3) according to its values on σ and on one of the characteristic sections AC and BC is called *Tricomi's problem.* If we, instead of AC (or BC), take a curve γ which is situated in the characteristic triangle ACB, monotonically going down and concave with respect to the axis $v = 0$, the problem is usually called a *general mixed problem.*

These stated problems are linear. The investigation of the problem of the existence of real motion on the physical plane encounters some difficulties in principle, even if these equations have solutions. To show the role of Tricomi's problem and its various generalizations when investigating the observed hydromechanical problem is still a topic for profound speculations.

Below we will mention some classes of nonlinear partial differential equations and the simplest classes of well-posed problems corresponding to them.

§5.2 Cauchy-Covalevski Equations

5.2.1. As in §2.1 of Chapter II we suppose that the equality

$$F(x, \dots, P_{i_1 \dots i_n}, \dots) = 0 \,, \tag{4}$$

where

$$P_{i_1 \dots i_n} = (p^1_{i_1 \dots i_n}, \dots, p^N_{i_1 \dots i_n}) = \frac{\partial^k u}{\partial x_1^{i_1} \dots \partial x_n^{i_n}} \,, \quad \sum_{j=1}^{n} i_j = k \,, \quad k = 0, \dots, m \,,$$

is a system of N partial differential equations with N unknown functions $(u_1, \dots, u_N) = u$, and the order of every equation equals $m, m \geq 1$.

It will be assumed, in addition, that the functions $F_i, i = 1, \dots, N$, are continuous in all their arguments in a neighbourhood of the fixed values x_0, $p^0_{i_1 \dots i_n}$, and are continuously differentiable with respect to $p_{m0\dots 0} = (p^1_{m0\dots 0}, \dots p^N_{m0\dots 0})$, and the function determinant

$$\det \left\| \frac{\partial F_l}{\partial p^j_{m0\dots 0}} \right\| \neq 0 \,, \quad j, l = 1, \dots, N \,. \tag{5}$$

Thus, by the well-known implicit function theorem we can write the system (4) in the form

$$\frac{\partial^m u}{\partial x_1^m} = f \left(x, \dots, \frac{\partial^k u}{\partial x_1^{l_1} \dots \partial x_n^{l_n}}, \dots \right) \,, \tag{6}$$

where the definite N-dimensional vector $f = (f_1, \dots f_N)$ is already independent of $\frac{\partial^m u}{\partial x_1^m}$.

The system (6) is usually known as the *Cauchy-Kovalevski system*. It is by no means evidently that each system (4) is of the Cauchy-Kovalevski type. For example, the condition (5) is not satisfied for the linear system

$$\frac{\partial u_1}{\partial x_1} = 0 \,, \quad \frac{\partial u_2}{\partial x_2} = 0 \,, \tag{7}$$

and hence it is not a Cauchy-Kovalevski system in the proposed notation. However, by using nonsingular replacement of independent variables $x_1 + x_2 = y_1, x_1 - x_2 = y_2$, the system (7) turns to the Cauchy-Kovalevski system

$$\frac{\partial v_1}{\partial y_1} = \frac{\partial v_1}{\partial y_2} \,, \quad \frac{\partial v_2}{\partial y_1} = \frac{\partial v_2}{\partial y_2} \,, \quad v_k = u_k \left(\frac{y_1 + y_2}{2} \,, \ \frac{y_1 - y_2}{2} \right) \tag{8}$$

in the variables y_1 and y_2

5.2.2. Suppose that there are smooth enough N-dimensional vectors

$$\varphi^{(k)}(x_2, \ldots, x_n) , \quad k = 0, \ldots, m-1 ,$$

given in a certain domain G of the hyperplane $x_1 = 0$ in the Euclidean space E_n of points x with Cartesian orthogonal coordinates $x_1, \ldots, x_n$.

The problem to find out, in some n-dimensional neighborhood G, a solution $u = (u_1, \ldots, u_N)$ of the system (6) which satisfies the condition

$$\frac{\partial^k u}{\partial x_1^k}\bigg|_{x_1 = cc} = \varphi^{(k)}(x_2, \ldots, x_n) , \quad k = 0, \ldots, m-1, c = \text{const.}, \tag{9}$$

is called a Cauchy's problem condition when the variable x_1 plays the role of time.

In Chapter I, the corresponding characteristic form $K(\lambda_1, \ldots, \lambda_n)$ of the system (4) was defined by the formula (2). We consider an $(n-1)$-dimensional surface S written in the form $\phi(x_1, \ldots, x_n) = 0$ in a domain D_0 of the space E_n where the system (4) is given. We will say, as in §4.1 of Chapter IV, that S is a characteristic surface for the system (4) if the equality $K(\phi_{x_1}, \ldots, \phi_{x_n}) = 0$ is satisfied at its every point x. It is clear that the data carrier given by $x_1 = c$ in the Cauchy problem (9) cannot be a characteristic surface for the system (6) of this paragraph, even if this system has real characteristic surfaces.

By introducing new unknown functions the system (6) can be reduced to a first order Cauchy-Kovalevski system as follows:

$$\begin{aligned} v_{x_1} &= f(x, v, p_{i_1 \ldots i_n}) , \\ p_{i_1 \ldots i_n} &= \left(p^1_{i_1 \ldots i_n}, \ldots, p^{N_1}_{i_1 \ldots i_n}\right) = \frac{\partial v}{\partial x_2^{i_2} \ldots \partial x_n^{i_n}} , \\ i_1 &= 0 , \quad \sum_{j=2}^{n} i_n = 1 , \end{aligned} \tag{10}$$

and the initial conditions (9) can be reduced to the initial conditions

$$v(0, x_2, \ldots, x_n) = \varphi(x_2, \ldots, x_n) , \tag{11}$$

where f and φ are given while v is an unknown N_1-dimensional vector, $N_1 > N$.

The system (10), furthermore, can be reduced to a form of quasilinear system. The situation is the same as for the case of $N = 1, m = 2, n = 2$, and we shall prove it by an example of a second order equation with two independent variables:

$$u_{x_1x_1} = f(x_1, x_2, u, u_{x_1}, u_{x_2}, u_{x_1x_2}, u_{x_2x_2}) , \tag{12}$$

$$u(0, x_2) = \varphi^{(0)}(x_2) , \quad u_{x_1}(0, x_2) = \varphi^{(1)}(x_2) . \tag{13}$$

We use the notations

$$u_1 = u(x_1, x_2) , \quad u_{1x_1} = u_2 , \quad u_{1x_2} = u_3 . \tag{14}$$

From the last two equalities in (14),

$$u_{2x_2} = u_{3x_1} . \tag{15}$$

From (12), (14) and (15) it follows that u_1, u_2 and u_3 are solutions of the Cauchy-Kovalevski system

$$\begin{aligned} &u_{1x_1} = u_2 , \quad u_{3x_1} = u_{2x_2} , \\ &u_{2x_1} = f(x_1, x_2, u_1, u_2, u_3, u_{2x_2}, u_{3x_2}) , \end{aligned} \tag{16}$$

with the initial conditions

$$\begin{aligned} &u_1(0, x_2) = u(0, x_2) = \varphi^{(0)}(x_2) , \\ &u_2(0, x_2) = u_{1x_1}|_{x_1=0} = u_{x_1}|_{x_1=0} = \varphi^{(1)}(x_2) , \\ &u_3(0, x_2) = u_{1x_2}|_{x_1=0} = u_{x_2}|_{x_1=0} = \frac{d}{dx_2}\varphi^{(0)}(x_2) , \end{aligned} \tag{17}$$

Introducing three new functions

$$u_4 = u_{2x_2} , \quad u_5 = u_{3x_2} , \quad u_6 = u_{2x_1} , \tag{18}$$

we come to the conclusion based on (15), (16) and (18) that the functions u_1, u_2, u_3, u_4, u_5 and u_6 must constitute a solution of the Cauchy-Kovalevski system

$$\begin{aligned} &u_{1x} = u_2 , \quad u_{2x_1} = u_6 , \quad u_{3x_1} = u_4 , \quad u_{4x_1} = u_{6x_2} , \\ &u_{5x_1} = u_{4x_2} , \quad u_{6x_1} = f_{x_1} + \sum_{l=1}^{5} f_{u_l} u_{lx_1} , \end{aligned} \tag{19}$$

where

$$u_6 = f(x_1, x_2, u_1, u_2, u_3, u_4, u_5) \ . \tag{20}$$

Due to (18) and (20) these functions should also satisfy the initial conditions

$$\begin{aligned} u_4(0, x_2) &= u_2(0, x_2)_{x_2} = \varphi^{(1)}_{x_2} \ , \\ u_5(0, x_2) &= u_3(0, x_2)_{x_2} = \varphi^{(0)}_{x_2 x_2} \\ u_6(0, x_2) &= f\left(0, x_0, \varphi^{(0)}, \varphi^{(1)}, \varphi^{(0)}_{x_2}, \varphi^{(1)}_{x_2}, \varphi^{(0)}_{x_2 x_2}\right) \ , \end{aligned} \tag{21}$$

in addition to (17).

Thus, for the quasilinear Cauchy-Kovalevski system (19) the problem (12)–(13) has been reduced to the Cauchy problem (17)–(21).

5.2.3. In the problem (12)–(13), as in the problems (16)–(17) and (19)–(17)–(21), the interval G of the straight line $x_1 = 0$ is the carrier of the initial data. When G is an interval of the straight line $x_1 = x_1^0$, then by a replacement of the independent variable, $x_1 = t + x_1^0$ it can be reached the carrier of the initial data will be the line $t = 0$. Furthermore, we can assume, without loss of generality, that the initial functions $\varphi^{(0)}$ and $\varphi^{(1)}$ in the conditions (13) are identically equal to zeros. Therefore we naturally consider the problem (6)–(8) stated as follows: to find a solution $u = (u_1, \ldots, u_N)$ of the Cauchy-Kovalevski system

$$u_t = f(x, t, u, u_{x_1}, \ldots, u_{x_n}) \ , \tag{22}$$

which satisfies the initial condition

$$u(x, 0) = 0 \tag{23}$$

where f is a real N-dimensional vector again given in the $(n+1)(N+1)$-dimensional domain $D_{(n+1)(N+1)}$ of the space of variables $x_1, \ldots, x_n, t, u$, $x_{x_1}, \ldots, u_{x_n}$, and u is a real N-dimensional vector in some $(n+1)$-dimensional neighbourhood D_{n+1} of the space E_{n+1} of variables $x_1, \ldots, x_n, t$. By D_n we denote a domain in the space E_n of variables $x_1, \ldots, x_n$.

In applications of the problem (22)–(23) the variable t often plays the role of time, and $x_1, \ldots, x_n$ that of spatial variables.

The problem (22)–(23) will be called a *Cauchy problem* with analytic initial data if near each point of the domain $D_{(n+1)(N+1)}$ the given vector f can be represented in the form of a sum of a power series with non-negative exponents:

$$f = \sum A_{kmsl_1\dots l_n}(x - x_0)^k(t - t_0)^m(u - u_0)^s(p_1 - p_1^0)^{l_i}\dots(p_n - p_n^0)^{l_n} ,$$

where

$$a_k = a_{k_1\dots k_n} , \quad \sum_{j=1}^{n} k_j = k , \quad (x - x_0)^k = (x - x_1^0)^{k_1}\dots(x_n - x_1^0)^{l_n} ,$$

$$A_{kmsl_1\dots l_n} = A_{k_1\dots k_n m_s\dots s_n, l_{1_1}\dots l_{n_1}\dots l_{n_N}} ,$$

$$\sum_{j=1}^{n} s_j = s , \quad \sum_{j=1}^{n} l_{i_j} = l_i , \quad i = 1,\dots,n .$$

The following important assertion, known as the Cauchy-Kovalevski theorem, holds: for every point $x^0 \in D_n$ there is a $(n+1)$-dimensional neighbourhood of variables (x,t) in which there exists one and only one analytic solution $u(x,t)$ of the Cauchy problem (22)–(23) with analytic initial data.

Some well-known concepts and facts in the theory of analytic functions of real variables are usually used to prove the Cauchy-Kovalevski theory.

Suppose that $f(y), y = (y_1,\dots,y_p)$ is an analytic function in some p-dimensional domain D_p of variables $y_1,\dots,y_p$. If $y^* = (y_1^*,\dots,y_p^*), y_j^* \neq y_j^0, j = 1,\dots,p$, is a point where the power series

$$f(y) = \sum a_k(y - y_0)^k , \tag{24}$$

with

$$a_k = a_{k_1\dots k_p} , \quad \sum_{j=1}^{p} k_j = k , \quad k = 0,1,\dots ,$$

$$y^0 = (y_1^0,\dots,y_p^0) , \quad (y - y^0)^k = (y_1 - y_1^0)^{k_1}\dots(y_p - y_p^0)^{k_p} ,$$

absolutely converges, then there exists a positive number M such that for all indices k the estimates

$$|a_k| \leq \frac{M}{\prod_{j=1}^{p}|y_j^* - y_j^0|^{k_j}} , \quad \sum_{j=1}^{p} k_j = k , \tag{25}$$

hold.

We say that a function $\varphi(y)$ which is analytic in D_p is a *majorant function* of the function $f(y)$ if all coefficients of the expansion

$$\varphi(y) = \sum b_k (y - y^0)^k \tag{26}$$

are positive and $|a_k| \leq b_k$ for all the values of indices k.

On the basis of the estimates (25) it follows that for a majorant of the function $f(y)$ represented by the formula (24) we can take the function

$$\varphi(y) = M \sum \frac{(y-y^0)^k}{\prod_{j=1}^p |y_j^* - y_j^0|^{k_j}} = \frac{M}{\prod_{j=1}^p \left(1 - \frac{y_j - y_j^0}{|y_j^* - y_j^0|}\right)} \tag{27}$$

when $|y_j - y_j^0| < |y_j^* - y_j^0|, j = 1, \ldots, p$.

The function

$$\varphi_1(y) = \frac{M}{1 - \frac{1}{a}\sum_{j=1}^p (y_j - y_j^0)} ,$$

where $a = \min_{1 \leq j \leq p} |y_j^* - y_j^0|$, is also a majorant of $f(y)$, in addition to (27).

The validity of this assertion follows from

$$\begin{aligned} \varphi_1(y) &= M \sum_{l=0}^{\infty} \frac{1}{a^l} \left\{ \sum_{j=1}^p (y_j - y_j^0) \right\}^l \\ &= M \sum_{l=0}^{\infty} \frac{1}{a^l} \sum \frac{l!}{k_1! \ldots k_p!} \prod_{s=1}^p (y_s - y_s^0)^{k_s} , \\ l &= \sum_{s=1}^p k_s , \end{aligned} \tag{28}$$

and

$$l! \geq \prod_{s=1}^p k_s! , \quad \frac{1}{a^l} \geq \frac{1}{\prod_{s=1}^p |y_s^* - y_s^0|^{k_s}} ,$$

if $\sum |y_j^* - y_j^0| < a$.

The function $\varphi_1(y)$ is evidently still a majorant of $f(y)$ if, instead of $y_1 - y_1^0$, in the expression (23), we write $\frac{1}{\alpha}(y_1 - y_1^0), 0 < \alpha < 1, \alpha = \text{constant}$.

When there are no spatial variables $x_1, \dots, x_n$, the problem (22)–(23) is the Cauchy problem

$$u(0) = 0 \tag{29}$$

for the ordinary differential equation

$$\frac{du}{dt} = f(t, u) \tag{30}$$

with an analytic $f(t, u)$ as its right-hand side.

In a neighbourhood of the point (0, 0) of the space E_2 of variables t and u we represent $f(t, u)$ in the form of a sum of the power series:

$$f(t, u) = \sum_{k,l=0}^{\infty} a_{kl} t^k u^l \ . \tag{31}$$

For simplicity of notation we assume that the convergence radii of the series (31), with respect to either t or u, are equal to one.

We now construct a function $u(x)$ which is analytic in a neighbourhood of the point $t = 0$, being a solution of the problem (24)–(30).

Assuming its existence (it will be found later) we can compute at the point $t = 0$, by sequentially differentiating the identity (30) and considering (29), the values of all the derivatives

$$n = 1, \dots, i + j \leq n - 1 \left(\frac{d^n u}{dt^n}\right)_0 = \omega_n(a_{00}, \dots, a_{ij}, \dots) \ , \tag{32}$$

where ω_n is a polynomial of its own arguments with positive coefficients. It is easy to see that the convergence radius of the power series

$$u(t) = \sum_{n=0}^{\infty} \frac{1}{n!} \left(\frac{d^n u}{dt^n}\right)_0 t^n \tag{33}$$

is different from zero.

In fact, on account of the convergence of the series (31) there exists a positive number M such that

$$a_{k,l} \leq \frac{M}{2} \ , \quad k, l = 0, 1, \dots \ . \tag{34}$$

Therefore, we can take the function

$$F(t,u) = \frac{M}{2(1-t)(1-u)} \tag{35}$$

for a majorant of $f(t,u)$.

By $U(t)$ we denote a solution of the ordinary differential equation

$$\frac{dU}{dt} = F(t,u) \tag{36}$$

with the initial condition

$$U(0) = 0 \ . \tag{37}$$

Applying the method of separation of variables we find the unique solution of the equation (36),

$$U = 1 - \sqrt{1 + M\log(1-t)} \ , \tag{38}$$

which satisfies the condition (37). Here $\log(1-t)$ is understood to be the branch of the function which vanishes when $t = 0$.

Since the expression $1 + M\log(1-t)$ vanishes when $t = 1 - e^{\frac{1}{M}}$ and its absolute value is less than one when

$$|t| < 1 - e^{-\frac{1}{M}} \ ,$$

the function expressed by the formula (38) is analytic in t and u, having a convergence radius at least equal to $1 - e^{-\frac{1}{M}}$. Thus, $U(t)$ is a majorant of $u(t)$, and hence it is possible to represent the latter in the form of its own Taylor series (33) of which the convergence radius is not less than $1 - e^{-\frac{1}{M}}$. Therefore the function $u(t)$, which is expressed in the form of a sum of the power series (33) with coefficients calculated by the formula (32), is a solution of the problem (29)–(30). It is evidently unique among all analytic solutions.

We shall now turn to the proof of the Cauchy-Kovalevski theory formulated above, and confine ourselves to considering the case when (22) is an equation with an unknown $u(x,t)$ and an analytic right-hand side f in a neighbourhood of the zero point of all arguments.

In the stated neighbourhood we have

$$\begin{gathered} f(x,t,u,p_1,\dots,p_n) = \sum a_{kmsl_1\dots l_n} x^k t^m u^s p_1^{l_1} \dots p_n^{l_n} \\ p_i = u_{x_i} \ , \quad i = 1,\dots,n \ . \end{gathered} \tag{39}$$

We assume, without loss of generality, that $a_{00\ldots0} = 0$. By replacing the unknown function by $u = a_{00\ldots0}t + v$ this can always be realized. In addition, we also assume that the convergence radii of the series (39) are all equal to one with respect to all variables, as we have done before.

For a majorant of the function f we can take the function

$$F = M\left\{\frac{1}{\left[1-\left(\frac{t}{\alpha}+u+\sum_{k=1}^{n}x_k\right)\right]\left(1-\sum_{k=1}^{n}p_k\right)}-1\right\}, \tag{40}$$

where M and α are some positive numbers and $0 < \alpha < 1$.

At the point $(x = 0, t = 0)$ we can compute the partial derivatives of all orders of the analytic solution $u(x,t)$ for the problem (22)–(23) and form the power series

$$u(x,t) = \sum b_{kl}x^k t^l \tag{41}$$

by differentiating term-by-term the identity (22) and taking into account (23).

The solution of the equation

$$\frac{\partial U}{\partial t} = F \tag{42}$$

with the initial condition

$$U(x,0) = 0 \tag{43}$$

is a majorant function of the function $u(x,t)$ given by the formula (41).

Near the point $(x = 0, t = 0)$ an analytic solution $v(x,t)$ of the equation (42) with positive coefficients expressed as the sum of a power series will be a majorant of the analytic solution $U(x,t)$ of the problem (42)–(43). We will seek a $v(x,t)$ in the form

$$v(x,t) = w(z), \tag{44}$$

where

$$z = \frac{t}{\alpha} + \sum_{k=1}^{n} x_k \tag{45}$$

Since we have by (44) and (45)

$$\frac{\partial v}{\partial t} = \frac{1}{\alpha}\frac{dw}{dz} , \quad \frac{\partial v}{\partial x_i} = \frac{dw}{dz} , \quad i = 1, \dots , n ,$$

we conclude, on the basis of (40) and (42), that the function $w(z)$ must be the solution of the ordinary differential equation

$$\frac{1}{\alpha}\frac{dw}{dz} = M \left[\frac{1}{(1-z-w)\left(1-n\frac{dw}{dz}\right)} - 1 \right] .$$

or equivalently,

$$\left(\frac{1}{\alpha} - Mn\right)\frac{dw}{dz} = \frac{n}{\alpha}\left(\frac{dw}{dz}\right)^2 + \frac{M}{1-z-w} - M . \tag{46}$$

Choosing the α such that the number $\frac{1}{\alpha} - Mn$ is positive, we can write the equation (46) in the form

$$\frac{dw}{dz} = \alpha^0 \left(\frac{dw}{dt}\right)^2 + \beta(z, w) , \tag{47}$$

where α^0 is a positive number and $\beta(z, w)$ is the sum of an absolutely convergent series with positive coefficients,

$$\beta(z, w) = \sum \beta_{kl} z^k w^l , \quad \beta_{00} = 0 . \tag{48}$$

From the two values of $\frac{dw}{dz}$ which satisfy the equality (47) we choose one which vanishes at $z = 0$ and $w = 0$. This means that $w(z)$ is a solution of the ordinary differential equation

$$\frac{dw}{dz} = \gamma(z, w) , \quad \gamma(0, 0) = 0 , \tag{49}$$

with the initial condition

$$w(0) = 0 . \tag{50}$$

An analytic solution of the problem (49)–(50), as we have already proved above, always exists and it is represented in the form of a sum of power series with a non-zero convergence radius:

$$w(z) = \sum_{k=2}^{\infty} \left(\frac{d^k w}{dz^k}\right)_0 \frac{z^k}{k!} . \tag{51}$$

It is easy to verify the coefficients of the power series are positive (51) by sequentially differentiating the equality (47) and considering (49) and (50).

Since the function w expressed by the formula (51) is a majorant of $U(x,t)$, and hence of $u(x,t)$, the power series (41) has positive convergence radii either in x or in t. Therefore, the sum of this series is an analytic solution of the problem (22)–(23) near the point $(x=0, t=0)$.

The proof of existence for an analytic solution in the small (near a given point) of the problem (22)–(23) in the case of one (scalar) equation with given analytic data offered here can be generalized immediately to the case of a system (with given analytic data); the uniqueness of analytic solution of this problem follows from the process of its construction. At the same time, the type of the system is immaterial. The essential thing is that the carrier $x_1 = 0$ of the initial condition (23) is not characteristic for the system (22). The straight line $x_1 =$ constant is a characteristic line for the decomposed system (7), and it cannot be the carrier of initial data as we already know (cf. §1.1 of Ch. I). As for the representation (8) of the same system, evidently, the straight line $y_1 = x_1 + x_2 =$ const is not a characteristic line.

The problem with the initial conditions $u(0,y) = 0$ and $u_x(0,y) = 0$ of the Monge-Ampere equation

$$u_{xx} \cdot u_{yy} - u_{xy}^2 = 0 \tag{52}$$

takes functions $u_1 = 0$ and $u_2 = x^2$ for solutions. This equation represented in the form

$$u_{xx} = \frac{u_{xy}^2}{u_{yy}}$$

can also be reduced to the type of Cauchy-Kovalevski equation, although its right-hand side is no longer an analytic function in u_{yy} along u_1 and u_2. For the same equation, the corresponding characteristic quadratic form of the equation (52), $u_{yy}dx^2 - 2u_{xy}dxdy + u_{xx}dy^2$, degenerates parabolically when $u = u_1$ and $u = u_2$.

The uniqueness of solution of the problem (22)–(23) can also be shown for the class of nonanalytic solutions. This problem, however, does not always have nonanalytic solution. It is easy to verify this by an example of the Cauchy-Riemann system

$$u_t = v_x\ , \quad v_t = -u_x\ , \quad x + it = z\ , \tag{53}$$

which is a Cauchy-Kovalevski system. In fact, we assume that in a domain D in the plane of variables x and t which contains a segment ab of the axis $t = 0$ in its interior there exists a regular solution $u(x,t), v(x,t)$ of the system (53) which satisfies the initial conditions

$$u(x,0) = \varphi(x) \ , \quad v(x,0) = 0 \ . \tag{54}$$

Taking into account the second of the conditions (54) we conclude on the basis of the *Riemann-Schwarz symmetry principle* that the function $u(x,t) + iv(x,t) = F(z)$ is analytic in the domain D which contains the open segment ab; that is, the function $\varphi(x)$ in the first of the conditions (54) has to be analytic with respect to variable x. Therefore, if we know in advance that $\varphi(x)$ does not have this property then the problem (53)–(54) will have no solution.

It is also easy to see that even analytic solutions of the problem (53)–(54) may be unstable. In fact, by direct check it is easy to verify that the functions

$$u_k = \frac{1}{k^2} \text{sh } kt \ \sin kx \ , \quad v_k = \frac{1}{k^2} \text{ch } kt \ \cos kx$$

are an analytic solution of the system (53) which satisfies the initial Cauchy's conditions

$$u_k(x,0) = 0 \ , \quad v_k(x,0) = \frac{1}{k^2} \cos \ kx \ , \tag{55}$$

but it does not satisfy the requirement for stability because if $k \to \infty$ in the second of the conditions (55) then we have $\lim_{k\to\infty} v_k(x,0) = 0$, while neither $u_k(x,t)$ nor $v_k(x,t)$ are even bounded when $k \to \infty$.

§5.3 Initial Value and Characteristic Cauchy Problems of Some Quasilinear Hyperbolic Equations

5.3.1. The following system is a special case of the Cauchy-Kovalevski system (22):

$$u_t + \lambda u_x = f(x,t,u) \ , \tag{56}$$

where $\lambda(x,t)$ is a given real and continuously differentiable diagonal matrix with elements $\lambda_k(x,t), k = 1, \ldots, N$, and the real vector $f = (f_1, \ldots, f_N)$ is given in certain simply-connected domain of independent variables x, t and for all values of the required vector $u = (u_1, \dots, u_N)$.

The system (56) is normally hyperbolic according to the definition given in §2.4 of Chapter II. For this we now consider the Cauchy problem

$$u(x,0) = 0 \ , \quad x \in I \ , \tag{57}$$

where the segment I of the straight line $t = 0$ is the carrier of the initial data.

This problem had been investigated in preceding paragraph under the assumption that f was on analytic vector with respect to all of its arguments.

In the plane of variables x and t the family of curves $C_k : x_k = x_k(t), k = 1, \dots, N$, defined by the ordinary differential equation

$$\frac{dx_k}{dt} = \lambda_k(x,t) \ , \quad k = 1, \dots, N \ , \tag{58}$$

is said to be the *family of characteristic curves* of the system (56). By E we denote a set of points (x,t) in the plane of the variables x and t which have the property that all the curves bundle of characteristics emanating from the point (x,t) and along the direction of the line $t = 0$ intersect with the segment I. By G we denote a domain which is contained in E and envelops I.

Since for a solution $u(x,t)$ of the system (56) the identity

$$\frac{du_k}{dt} = \frac{\partial u_k}{\partial t} + \lambda_k(x,t)\frac{\partial u_k}{\partial x} \ , \quad k = 1, \dots, N \ ,$$

holds along the curves C_k from the point $[\xi_k(x,t), 0]$ to $(x,t) \in G$ due to (58), we can give the equality (56) the form

$$u(x,t) = \int_{(\xi_k,0)}^{(x,t)} f(\xi_k, \tau, u)_{\xi_k = \xi_k(\tau)} d\tau \tag{59}$$

on account of (57) where $[\xi_k(x,t), 0]$ is the intersection point of the characteristic curve $\xi_k = \xi_k(\tau)$ emanated from the point (x,t) with the segment I.

Corresponding to each vector $u(x,t)$ which is continuously differentiable in the domain G and satisfies the condition (57), the integral operator T on the right-hand side of the formula (59) sets up a vector $v(x,t)$ which is continuously differentiable in the same domain and satisfies the same condition.

The problem (56)–(57) will be solved if we succeed in finding a fixed point for the transformation $v = Tu$, namely, the vector $u(x,t)$ for which the equation

$$u = Tu \tag{60}$$

holds.

A solution of the equation (60) can be constructed by a sequential approximation method. Let as require furthermore the continuity in x and t and the continuous differentiability in $u_1, \ldots, u_N$ for all the components of the vector f, and the finite upper bound M for the absolute values of $\frac{\partial f}{\partial u_k}$, $k = 1, 2, \ldots, N$. For the zero-th approximation solution of the equation (60) we take the vector $u^{(0)} = 0$, and the subsequent approximations are defined by the formulae

$$u^{(n)} = Tu^{(n-1)} .$$

It is assumed that $t < h$, where h is a positive constant.

For the difference $u^{(2)} - u^{(1)}$ we have

$$\begin{aligned} |u^{(2)}(x,t) - u^{(1)}(x,t)| &= \left| \int_{(\xi,0)}^{(x,t)} [f(\xi,\tau,u^{(1)} - f(\xi,\tau,0)] d\tau \right| \\ &\le \int_{(\xi,0)}^{(x,t)} \left| \sum_{k=1}^{N} \tilde{f}_k^{(1)} u_k^{(1)} \right| d\tau \end{aligned}$$

by the finite increment theorem, where $\tilde{f}_k^{(1)}$ are mean values of $\frac{\partial f}{\partial u_k^{(1)}}$.

By repeating this argument we find that for each natural number n the estimate

$$\left| u^{(n+1)}(x,t) - u^{(n)}(x,t) \right| \le \theta \left| u^{(n)}(x,t) - u^{(n-1)}(x,t) \right| \quad n = 1, 2, \ldots , \tag{61}$$

where $\theta = MNh$, holds. It can be assumed for sufficiently small h that $\theta < 1$. Therefore the sequence $u^{(n)}, n = 1, 2, \ldots,$ converges uniformly and its limit $u(x,t)$ is the solution of the equation (60). It can be proved that $u(x,t)$ has continuous derivatives in x and t, namely, $u(x,t)$ is a solution of the problem (56)–(57). Uniqueness of the solution is obvious.

By using the notations $u_x = u_1$ and $u_t = u_2$ the Cauchy problem

$$u(x,0) = 0 \ , \quad u_t = (x,0) = 0$$

of the quasilinear hyperbolic type equation

$$u_{xx} - u_{tt} = f(x, t, u_x, u_t)$$

is reduced to a system of the form (56):

$$u_{1t} - u_{2x} = 0 , \quad u_{2t} - u_{1x} = f(x, t, u_1, u_2) .$$

with the initial conditions $u_1(x, 0) = 0$ and $u_2(x, 0) = 0$. Therefore, its solution exists and is unique at least in the small scope.

5.3.2. We now consider the equation

$$u_{xy} = f(x, y, u, p, q) , \tag{62}$$

where f is a given function of $x, y, u, p = u_x$ and $q = u_y$.

The lines $x =$ constant and $y =$ constant are the characteristic lines of the equation (62). These lines are evidently not suitable as carriers of the Cauchy initial data for the equation (62), but they can be used as carriers of initial data of the Cauchy characteristic problem, or as we usually say, of the Goursat problem

By D we denote the rectangular domain in the plane of variables x and y bounded by the lines $x = 0, y = 0, x = a$ and $y = b$. We suppose that the function f on the right-hand side is subject to the requirements: (1) it is continuous in all of its arguments for $(x, y) \in D$ and

$$|u| < A , \quad |p| < A_2 , \quad |q| < A_3 , \tag{63}$$

where A_1, A_2 and A_3 are positive, (2) it satisfies the Lipschitz's condition with respect to u, p and q, namely, there are positive numbers k_1, k_2 and k_3 such that for arbitrary $u', p', q', u'', p'', q''$ satisfying the conditions (63) the estimate

$$|f(x, y, u', p', q') - f(x, y, u'', p'', q'')| \leq k_1|u' - u''| + k_2|p' - p''| + k_3|q' - q''| \tag{64}$$

holds if $(x, y) \in D$.

By a Goursat problem for the equation (62) we mean the requirement to determine its solution which is continuous in $D \cup \partial D$ and regular in D when the values $u(x, 0) = \varphi(x)$ for $0 \leq x \leq a, u(0, y) = \psi(y)$ for $0 \leq y \leq b$ and $\varphi(0) = \psi(0)$ are given in advance.

When the functions φ and ψ are continuously differentiable we obtain the conditions of the Goursat problem

$$\begin{aligned} u(x,0) &= 0 \quad \text{for} \quad a \le x \le a \, , \\ u(0,y) &= 0 \quad \text{for} \quad 0 \le y \le b \end{aligned} \tag{65}$$

by replacing $u(x,y)$ with $u(x,y) - \varphi(x) - \psi(y) + \varphi(0)$.

Again, we use a sequential approximation method to obtain the solution of the problem (62)–(65). We take $u_0(x,y) = 0$ for the zero-th approximation, then for subsequent approximations we have

$$u_k(x,y) = \int_0^x d\xi \int_0^y f\left(\xi, \eta, u_{n-1}, \frac{\partial u_{n-1}}{\partial \xi}, \frac{\partial u_{n-1}}{\partial \eta}\right) d\eta \, , \quad h = 1, 2, \ldots \, . \tag{66}$$

By α and β we denote two positive numbers of not more than a, b respectively, and also

$$\alpha\beta < \frac{A_1}{M} \, , \quad \alpha < \frac{A_3}{M} \, , \quad \beta < \frac{A_2}{M} \, , \tag{67}$$

where M is the maximum of f while its argument is subject to the requirement (63).

It is obvious from (67) that when $(x,y) \in D_1$ and $D_1 = \{0 \le x \le \alpha, 0 \le y \le \beta\}$ the values defined by the formula (66),

$$u_n(x,y) \, , \quad \frac{\partial u_n(x,y)}{\partial x} \, , \quad \frac{\partial u_n(x,y)}{\partial y} \, , \quad n = 1, 2, \ldots \, ,$$

satisfy the conditions (63).

We will now show that the sum $u(x,y)$ of the series

$$u_1(x,y) + \sum_{n=2}^{\infty} [u_n(x,y) - u_{n-1}(x,y)] \tag{68}$$

is the solution of the problem (63)–(65) if f satisfies the requirements (1) and (2).

We adopt the notations

$$A = \max(A_1, A_2, A_3) \, , \quad k = \max(k_1, k_2, k_3) \, , \quad N = \max\left[3A, A\left(2 + \frac{x+y}{n}\right)\right] \, .$$

By virtue of (66) we have from the inequality (67)

$$|u_1(x,y)| \leq A \ , \quad |u_{1x}| \leq A \ , \quad |u_{1y}| \leq A \ . \tag{69}$$

Furthermore, by making use of (64) we obtain on the basis of (66) and (69) that

$$|u_2 - u_1| \leq \int_0^x d\xi \int_0^y (k_1|u_1| + k_2|u_{1\xi}| + k_3|u_{1\eta}|)d\eta \leq kN\frac{(x+y)^2}{2!} \ ,$$
$$|(u_2 - u_1)_x| \leq kN(x+y) \ , \quad |(u_2 - u_1)_y| \leq kN(x+y) \ .$$

Continuing this process we find that for every positive integer n,

$$\begin{aligned} |u_n - u_{n-1}| &\leq k^{n-1}N^{n-1}N\frac{(x+y)^n}{n!} \ , \\ |(u_n - u_{n-1})_x| &\leq k^{n-1}N^{n-1}\frac{(x+y)^{n-1}}{(n-1)!} \ , \\ |(u_n - u_{n-1})_y| &\leq k^{n-1}N^{n-1}\frac{(x+y)^{n-1}}{(n-1)!} \ . \end{aligned} \tag{70}$$

The conclusion based on the estimates (70) is that the series (68) converges absolutely and uniformly and it is differentiable in x and y term by term. It is clear that the sum of this series $u(x,y) = \lim_{n\to\infty} u_n(x,y)$. On the basis of this and by the limiting process we conclude that $u(x,y)$ is a solution of the problem (62)–(65). This problem cannot have more than one solution. In fact, supposing it has two solutions

$$u(x,y) = \int_0^x d\xi \int_0^y f(\xi,\eta,u,u_\xi,u_\eta)d\eta \ ,$$
$$v(x,y) = \int_0^x d\xi \int_0^y f(\xi,\eta,v,v_\xi,v_\eta)d\eta \ ,$$

for the difference $w = u - v$ we have

$$w(x,y) = \int_0^x d\xi \int_0^y [f(\xi,\eta,u,u_\xi,u_\eta) - f(\xi,\eta,v,v_\xi,v_\eta)]\,d\eta \ . \tag{71}$$

Hence by virtue of (64) we have.

$$|w| \leq kN\frac{(x+y)^2}{2!} \ , \quad |w_x| \leq kN(x+y) \ , \quad |w_y| \leq kN(x+y) \tag{72}$$

where k and N are the positive constants introduced above.

On the basis of (72), from (71) we obtain again

$$|w| \leq k^2 N^2 \frac{(x+y)^3}{3!} , \quad |w_x| \leq k^2 N^2 \frac{(x+y)^2}{2!} , \quad |w_y| \leq k^2 N^2 \frac{(x+y)^2}{2!} .$$

Continuing this process we come to the conclusion that for arbitrary positive integer n

$$|w(x,y)| \leq k^n N^n \frac{(x+y)^{n+1}}{(n+1)!} .$$

From this it follows that $w(x,y) = 0$, namely, $u(x,y) = v(x,y)$.

The uniqueness of solution of the problem (62)–(65) may not be true if the requirement (64) is violated. This is easily verified by the example

$$u_{xy} = u^{\frac{1}{2}} , \tag{73}$$

which has regular solutions $u_1 = 0$ and $u_2 = \frac{1}{16}x^2y^2$ satisfying the conditions (65). The reason for this fact must be that the right-hand side f of the equation (73) does not satisfy the requirement (64).

In addition to the remark, the Monge-Ampere equation

$$u_{xx}u_{yy} - u_{xy}^2 = -1 , \tag{74}$$

which is hyperbolic along its any arbitrary solution, may have non-unique solution for the problem (55). The hyperbolicity of the equation (74) follows from the positive definiteness of the discriminant which corresponds to its characteristic form $Q = u_{yy}dy^2 + 2u_{xy}dxdy + u_{xx}dx^2$ along an arbitrary solution $u(x,y)$. Particularly, along the solutions $u_1 = xy$ and $u_2 = -xy$ it has $Q = \pm dxdy$. Therefore, the lines $x = 0$ and $y = 0$ are characteristic lines of the equation (74) along its solutions u_1 and $-u_1$ where they vanish. This means that the Goursat problem (65) of the equation (74) is not well-posed. The equation (74), of course, does not belong to the class of equations of the form (62).

§5.4 Dirichlet's Problem of Nonlinear Elliptic Equations

5.4.1. Suppose that the second order nonlinear partial differential equation

$$F(x, \ldots, p_{i_1 \ldots i_n}) = 0 \tag{75}$$

$$x = (x_1, \ldots, x_n) , \quad p_{i_1 \ldots i_n} = \frac{\partial^k u}{\partial x_1^{i_1} \ldots \partial x_n^{i_n}} , \quad \sum_{j=1}^{n} i_j = k , \quad k = 0, 1, 2,$$

is given for all values of the independent variables $x_1, \dots, x_n$ in the domain D of the Euclidean space E_n of points x and for all values of $p_{i_1 \dots i_n}$

We assume that the equation (75) is uniformly elliptic, namely, there are constants k_0 and k_1 with the same sign such that for all values of x and $p_{i_1 \dots i_n}$ the estimate

$$k_0 \sum_{i=1}^{n} \lambda_i^2 \leq Q(\lambda_1, \dots, \lambda_n) \leq k_1 \sum_{i=1}^{n} \lambda_i^2 \,. \tag{76}$$

holds, where

$$Q(\lambda_1, \dots, \lambda_n) = \sum \frac{\partial F}{\partial p_{i_1 \dots i_n}} \lambda_1^{i_1} \dots \lambda_n^{i_n} \,, \quad \sum_{j=1}^{n} i_j = 2 \,, \tag{77}$$

and, furthermore,

$$\frac{\partial F}{\partial u} \leq 0 \tag{78}$$

if Q is a positive definite form, and

$$\frac{\partial F}{\partial u} \geq 0$$

if Q is a negative definite form.

It is not difficult to see that the Dirichlet problem

$$u(x) = \varphi(x) \,, \quad x \in S \,, \quad u(x) \in \mathbf{C}^{0,0}(D \cup S) \tag{79}$$

in a bounded domain D with a $(n-1)$-dimensional boundary S has no more than one solution.

In fact, by the finite increment theorem, for the difference $u(x)$ of two solutions u_1 and u_2 for the problem (75)–(79) we have

$$F(x, \dots, p'_{i_1 \dots i_n}, \dots) - F(x, \dots, p^2_{i_1 \dots i_n}, \dots) = \sum \tilde{f}_{i_1 \dots i_n} \frac{\partial^k u}{\partial x_1^{i_1} \dots \partial x_n^{i_n}} \,,$$

where

$$p^m_{i_1 \dots i_n} = \frac{\partial^k u_m}{\partial x_1^{i_1} \dots \partial x_n^{i_n}} \,, \quad m = 1, 2,$$

and $\tilde{f}_{i_1\ldots i_n}$ is that mean value of the functions

$$f_{i_1\ldots i_n} = \frac{\partial F}{\partial p_{i_1\ldots i_n}} \,, \quad \sum_{j=1}^{n} i_j = k \,, \quad k = 0,1,2 \,. \tag{80}$$

Thus, $u(x)$ identically satisfies the equality

$$\sum \tilde{f}_{i_1\ldots i_n} \frac{\partial^k u}{\partial x_1^{i_1} \ldots \partial x_n^{i_n}} = 0 \,, \quad \sum_{j=1}^{n} i_j = k \,, \quad k = 0,1,2 \,. \tag{81}$$

Repeating the reasoning introduced in §3.8 of Chapter III when proving the uniqueness of solution for the Dirichlet problem of second order uniformly elliptic linear equation and considering (78), (79), (80) and (81), one can see that $u(x) = 0$, that is, $u_1(x) = u_2(x)$ in the whole domain D.

Note that we should be careful when applying the stated conclusion of the uniqueness of solution for the problem (75)–(79). For example, the Monge-Ampere equation

$$u_{xx}u_{yy} - u_{xy}^2 = 4 \tag{82}$$

is elliptic along its arbitrary solution because the discriminant which corresponds to its characteristic form (77) is equal to -4. This form is

$$Q(dx, dy) = u_{kyy}dy^2 + u_{kxx}dx^2 \,, \quad i = 1,2 \,, \tag{83}$$

along the solutions $u_1 = x^2 + y^2 - 1$ and $u_2 = -u_1$, and, hence, it is positively definite along u_1 and negatively definite along u_2. Nevertheless, the equation (82) in the circle $D : \{x^2 + y^2 < 1\}$ has $u_1(x,y)$ and $-u_1(x,y)$ as its regular solutions which vanish on the circumference $S : \{x^2 + y^2 = 1\}$. In the case under consideration the form (79) does not keep its sign along all the possible solutions $u(x,y)$ of the equation (82).

5.4.2. The condition (79) can be made homogeneous if its right-hand side is smooth enough. In the case of the equation

$$\Delta u = f(x,y,u) \,, \quad \Delta = \frac{\partial^2}{\partial x^2} + \frac{\partial^2}{\partial y^2} \,, \tag{84}$$

Dirichlet's problem

$$u(x,y) = 0 \ , \quad (x,y) \in S \tag{85}$$

in a domain with a sufficiently smooth boundary is immediately reduced to the second kind nonlinear Fredholm's integral equation

$$u(x,y) + \frac{1}{2\pi} \int_D G(x,y;\xi,\eta) f(\xi,\eta,u) d\xi d\eta = 0 \ , \tag{86}$$

where $G(x,y,\xi,\eta)$ is a Green's function of the Dirichlet problem for harmonic functions in the domain D.

If we express the sequential approximations by

$$u_0(x,y) = 0 \ , \quad u_n(x,y) = -\frac{1}{2\pi} \int_G G(x,y;\xi,\eta) f(\xi,\eta,u_{n-1}) d\xi d\eta \ , \tag{87}$$

and assume the boundedness of $|f_u|$, then it can be shown that a limit of the sequence exists when the measure of the domain D is small enough, and it gives a solution of the problem (84)–(85). When, furthermore, we know that $f_u \geq 0$, then it can be proved that the sequence (87) converges uniformly in D and its limit $u(x,y)$ is a desired solution of this problem provided that the domain D is bounded.

§5.5 Cauchy's Problem for a Class of First Order Quasi-linear Equations

5.5.1. In this paragraph we shall consider a first order quasi-linear equation with the form

$$a \nabla u = 0 \ , \tag{88}$$

where ∇ is the gradient operator in variables $x_1, \dots, x_n$ and $a(u) = [a_1(u), \dots, a_n(u)]$ is a given non-zero n-dimensional and continuously differentiable vector defined for all values of the desired solution $u(x), x = (x_1, \dots, x_n)$.

Let $\alpha(u) = [\alpha_1(u), \dots, \alpha_n(u)]$ be an arbitrary continuously differentiable vector which is orthogonal to the vector $a(u)$,

$$a(u)\alpha(u) = 0 \ . \tag{89}$$

It is easy to see that if a continuously differentiable function $g(\alpha x)$ of the scalar argument $\alpha x = \sum_{i=1}^{n} \alpha_i x_i$ satisfies the condition

$$g'(\alpha x)(\alpha' x) \neq 1 , \tag{90}$$

then the function $u(x)$ defined by the equality

$$u - g(\alpha x) = 0 \tag{91}$$

is the solution of the equation (88).

In fact, due to (90) there is an implicit function $u(x)$ defined by the equality (91) and

$$\nabla u = \frac{g'(\alpha x)\alpha}{1 - g'(\alpha x)(\alpha' x)} . \tag{92}$$

By substituting the expression ∇u given by (92) in the left-hand side of (88), we can see the validity of the stated conclusion on account of (89).

5.5.2. In particular, the function $u(x,t)$ defined by the equality

$$u - g(at + bx) = 0 , \tag{93}$$

where $g(\eta), \eta = at + bx$, is an arbitrary continuously differentiable function satisfying the condition

$$(a't + b'x)g' \neq 1 ,$$

and $a(u), b(u)$ are given continuously differential functions, is the solution of the equation

$$a(u)u_x - b(u)u_t = 0 , \quad a^2 + b^2 \neq 0 . \tag{94}$$

The equation

$$u_t + a(u)u_x = 0 , \tag{95}$$

which is well known in applications as the kinematic wave equation, is a special case of (94) when $b = -1$.

By virtue of the most general solution of equation (95) is obtained from the formula (96)

$$u = F[x - a(u)t] , \tag{96}$$

where F is an arbitrary continuously differentiable function satisfying the condition

$$F'(\omega)a'(u)t \neq -1 , \quad \omega = x - a(u)t .$$

The formula (96) allows us to investigate exhaustively Cauchy's problem formulated as follows: to seek a regular solution $u(x,t)$ of the equation (95) from the initial condition

$$u(x,0) = u_0(x) , \quad -\infty < x < \infty . \tag{97}$$

where $u_0(x)$ is a given continuously differentiable function.

Requiring that the function $u(x,t)$ defined by the formula (96) satisfy the initial condition (97), we find that for all values of $x, -\infty < x < \infty$,

$$F(x) = u_0(x) ,$$

and, therefore,

$$u(x,t) = u_0(x - at) . \tag{98}$$

The formula (98) allows us especially to see clearly a situation of non-uniqueness and singularity for the solutions of the problem (95)–(97). Particularly, if $a(u) = u, u_0 = x$ then the function

$$u(x,t) = \frac{x}{t+1}$$

is the solution of this problem on the whole plane of the variables x and t, except for $t = -1$ along which it has a discontinuity.

Lowering the order of smoothness for functions $a(u)$ and $u_0(x)$ may lead to breakdown of uniqueness and appearance of discontinuities for solutions for the problem (95)–(97).

For clearness let us turn our attention to two examples.

At first we suppose that $a(u) = u, u_0(x) = \text{sgn } x, -\infty < x < \infty$. From (96)–(97) we have

$$u = \text{sgn}(x - ut) . \tag{99}$$

Thus, from an ordinary differential equation of characteristics which correspond to the partial differential equation (95)

$$dx - udt = 0 , \tag{100}$$

it follows that along the solution $u(x, t)$ obtained from (99) of the problem (95)–(97) there is a characteristic straight line passing through each point $(x_0, 0)$, which is $x + t = x_0$ if $x_0 < 0$, or $x - t = x_0$ if $x_0 > 0$. The lines $x + t = 0$ and $x - t = 0$ divide each neighbourhood D of the point (0, 0) into four parts which are denoted by $D^{\text{sgn}(t+x)}_{\text{sgn}(t-x)}$. It is clear that in D^+_- and D^-_+ there exist solutions $u = 1$ and $u = -1$ of the problem (95)–(97) respectively, and in D^+_+ there is no solution, but in D^-_- there are two solutions $u_1 = 1$ and $u_2 = -1$.

Now suppose $a(u) = u, u_0(x) = x\text{sgn } x, -\infty < x < \infty$. Because of (98),

$$u(x, t) = \omega \text{sgn} \omega \ , \quad \omega = x - ut \ ,$$

the solutions of the problem (95)–(97),

$$u_1 = x - u_1 t \ , \quad u_2 = -x + u_2 t \ ,$$

are obtained immediately for the case, namely,

$$u_1 = \frac{x}{1+t} \ , \quad u_2 = \frac{x}{t-1} \ , \tag{101}$$

Replacing the u in the left-hand side of the equation (100) with the obtained expressions (101) we can see that the characteristic straight lines $x = c(t+1)$ and $x = c(t-1)$, where $c = \text{constant} > 0$, correspond to the solutions u_1 and u_2 of the problem (95)–(97). Therefore, in the zone between the lines $t = 1$ and $t = -1$ there is a continuous function

$$u = \begin{cases} \dfrac{x}{1+t} \ , & x \geq 0 \ , \\[2ex] \dfrac{x}{t-1} \ , & x \leq 0 \ , \end{cases}$$

which satisfies the initial condition $u(x, 0) = x\text{sgn}x$ and is the solution of the equation

$$u_t + uu_x = 0 \tag{102}$$

either in the interior of the right half-zone $x > 0$ or in the interior of the left half-zone $x < 0$. There is a straight line in either the family of the characteristic lines $x = c(t+1)$ or that of $x = c(t-1)$ through every point either in the part D_1 in the half-plane $x < 0$ and above the straight line $t = 1$ or the part D_2 in the half-plane $x < 0$ and under the straight line $t = -1$. But there is no straight line in these families through either the part D_3 in the half-plane $x < 0$ and above the straight line $t = 1$ or the part D_4 in the half-plane $x > 0$ and under the straight line $t = -1$. Thus, in the domains D_1 and D_2 there are characteristic line preceeded from D that we just mentioned in accordance with the two solutions of the equation (102), whereas in D_3 and D_4 the solution from D does not extend.

§5.6 Linearization of a Class of Quasilinear Equations

5.6.1. By some transformation of the required function

$$u = \omega(v) \ , \tag{103}$$

where $\omega(v)$ and $v(x)$ are new unknown functions, the quasilinear equation

$$\sum_{i,j=1}^{n} a^{ij}(a)\left[u_{x_i x_j} - b(u)u_{x_i}u_{x_j}\right] + \sum_{i=1}^{n} c^i(x)u_{x_i} + d(x,u) = 0 \ , \tag{104}$$

where $a^{ij}(x), b(u), c^i(x)$ and $d(x, u)$ are given functions, can have the form

$$\sum_{i,j=1}^{n} a^{ij}\left[\omega'' - b(\omega)\omega'^2\right] v_{x_i}v_{x_j} + \omega'\left(\sum_{i,j=1}^{n} a_{ij}v_{x_i x_j} + \sum_{i=1}^{n} c^i v_{x_i}\right) + d = 0 \ ,$$

$$\omega' = \frac{d\omega}{dv} \ , \quad \omega'' = \frac{d^2\omega}{dv^2} \ . \tag{105}$$

By choosing $\omega(v)$ as a solution of the second order ordinary nonlinear differential equation

$$\omega'' - b(\omega)\omega'^2 = 0 \ , \tag{106}$$

for determining the function $v(x)$ we obtain from (105), a partial differential equation which is linear with respect to its second and first order derivatives:

$$\sum_{i,j=1}^{n} a^{ij}(x)v_{x_i x_j} + \sum_{i=1}^{n} c^i(x)v_{x_i} + \frac{1}{\omega'}d(x,u) = 0 . \tag{107}$$

Thus, if the functions $\omega(x)$ and $v(x)$ are the corresponding solutions of the equations (106) and (107) respectively, then the function $u(x)$ defined by (103) will be a solution of the equation (104).

The solution of the equation (106) can be expressed as a quadrature

$$v = \alpha \int_0^u \exp(-\int_0^\tau b(t)dt)d\tau + \beta , \tag{108}$$

where α and β are arbitrary constants. As to the equation (107) in the case when

$$\frac{1}{\omega'}d[x,\omega(v)] = d_1(x)v + d_2(x) ,$$

it is linear

$$\sum_{i,j=1}^{i} a^{ij}(x)v_{x_i x_j} + \sum_{i=1}^{n} c^i(x)v_{x_i} + d_1(x)v + d_2(x) = 0 . \tag{109}$$

In a series of cases, boundary, initial and other problems, posed for the equation (104) depending on the type, generate the corresponding problems for the equation (109) by the formula (108), and, moreover, from the equality (108), u can be uniquely determined as a function of v. Then the original problem is obviously well-posed. It should be singled out that research into bifurcation of solutions of the equation (103) can be reduced to research into the Riemann surface of the functional dependence (108) between u and v.

5.6.2. As a first example of equations with the form (103) we consider the equation

$$\square\, u + \frac{1}{2}\mu^2 u \,\square\, u^2 = 0 , \tag{110}$$

where μ is a real constant and $\square$ is the *d'Alembertian*

$$\square = \sum_{i,j=1}^{n} a^{ij}(x)\frac{\partial^2}{\partial x_i \partial x_j}$$

in the 4-dimensional space of variables x_1, x_2, x_3 and x_4 with the linear element

$$ds^2 = \sum_{i,j=1}^{n} g_{ij}(x)dx_i dx_j \ ,$$

and $u(x_1, x_2, x_3, x_4)$ is the required real function of the spatial variables x_1, x_2, x_3 and time $t = x_4$.

$$(1 + \mu^2\omega^2)\omega'' + \mu^2\omega\omega'^2 = 0 \tag{111}$$

and

$$\Box \ v = 0 \tag{112}$$

we can easily draw the conclusion that the most general solution of (110) can be written in the implicit form

Since the equations (106) and (109) in the considered case have the form

$$\varphi(u, v) = 0 \ , \tag{113}$$

where

$$\varphi = \omega\sqrt{1 + \mu^2\omega^2} + \frac{1}{\mu}\text{arc sh } \mu\omega - v \ , \tag{114}$$

$u = \omega(v)$ is a general solution of the equation (111) and $v(x_1, x_2, x_3, t)$ is a general solution of the equation (112).

Cauchy's problem for the equation (110) with arbitrary sufficiently smooth initial data

$$u(x, t_0) = \tau(x) \ , \quad u_t(x, t_0) = \nu(x) \ , \quad x = (x_1, x_2, x_3) \ , \tag{115}$$

corresponds to Cauchy's problem for the equation (112) with the initial data

$$\begin{aligned} v(x, t_0) &= \tau\sqrt{1 + \mu^2\tau^2} + \frac{1}{\mu}\text{arc sh } \mu\tau \ , \\ v_t(x, t_0) &= z\nu\sqrt{1 + \mu^2\tau^2} \ . \end{aligned} \tag{116}$$

Writing the equation (110) in the form

$$\Box \ u + \frac{\mu^2 u}{1 + \mu^2 u^2} \sum_{i,j=1}^{n} a^{ij} u_{x_i} u_{x_j} = 0$$

and considering real variables μ and u we can make sure that the equation is of Cauchy-Kovalevski type if the operator is hyperbolic, and conclude that the problem (112)–(116) is well-posed. From this and taking into account the situation that the sufficient condition for existence of u as an implicit function of v is

$$\varphi_\omega = 2\sqrt{1+\mu^2\omega^2} \neq 0 .$$

we arrive at the representability of all regular solutions u of the equation (110) through the formula (113). This conclusion is still valid when $\Box$ is an elliptic operator with analytic coefficients if the initial data (115) are analytic. Therefore the problem (110)–(115) always has one and only one solution in all cases under these assumptions.

We will now consider the elliptic type equation

$$\Delta u - \frac{1}{u}(\nabla u)^2 = 0 , \tag{117}$$

where ∇ is the gredient operator in variables $x_1, \dots , x_n$ and $\Delta = \nabla\nabla$.

By virtue of (108) the relation between functions u and v is now given by

$$u = e^v , \tag{118}$$

where v is, generally speaking, an arbitrary complex harmonic function of variables $x_1, \dots , x_n$.

Let us consider the Dirichlet problem in a bounded domain D with an $(n-1)$-dimensional boundary $S = \partial D$,

$$u(x) = f(x) , \quad x \in S , \tag{119}$$

where $f(x)$ is a given real continuous function.

On the basis of the formula (118) we conclude that the problem (117)–(119) always has one and only one regular solution in the domain D which is of constant signs (and we can assume that it is positive without loss of generality) and continuous on $D \cup S$.

The solution of the problem (117)–(119) might not be unique if we waived the requirement of constant signs for the required solution $u(x)$.

In fact, let us observe the case $n = 2$, when the domain D is the circle $|x| < 1$ and $f(x) = 1$ on the circumference $S : |x| = 1$. It follows from the formula (118) that the function $u_1 \cdot u_2$ is also a solution of the equation (117) in addition to $u_1(x)$ and $u_2(x)$. Therefore it is clear that the family of functions

$$u(x) = |F(z)|^2 \exp\left(-\frac{1}{\pi}\int_S \frac{1-|z|^2}{|t-z|^2}\log|F(t)|ds\right) \ , \quad t = e^{is} \ ,$$

where $F(z)$ is an arbitrary function of complex variable $z = x_1 + ix_2$, which is analytic in D, continuous in $D \cup S$ and different from zero on S, are also solutions of the problem (117)–(119) for the considered case.

Note that the equation (117) is relative to the kind of equations discussed in §5.4 of this chapter though it is not subject to the condition (78) and the left-hand side of this formula is no longer bounded when $u \to 0$.

The hyperbolic-type equation

$$u_{x_1x_1} - u_{x_2x_2} - \frac{1}{u}\left(u_{x_1}^2 - u_{x_2}^2\right) = 0 \tag{120}$$

is obviously of the form (104).

The solutions of this equation can be represented by the formula

$$u(x_1, x_2) = f_1(x_1 + x_2)f_2(x_1 - x_2) \ , \tag{121}$$

where f_1 and f_2 are arbitrary, twice continuously differentiable functions.

In the triangle $A(-1,0)B(1,0)C(0,\frac{1}{2})$ in the plane of Cartesian orthogonal coordinates x_1 and x_2 there is a solution $u(x_1, x_2)$ of constant signs for the Cauchy problem

$$u(x_1, 0) = \tau(x_1) \ , \quad u_{x_2}(x_1, 0) = \nu(x_1) \ , \quad -1 < x < 1 \ , \tag{122}$$

of the equation (120), which due to (121) is given by the formula

$$u(x_1, x_2) = \sqrt{\tau(x_1 + x_2)\tau(x_1 - x_2)} \exp\left[\frac{1}{2}\int_{x_1 - x_2}^{x_1 + x_2} \frac{\nu(t)}{\tau(t)}dt\right]$$

and is unique.

The problem (120)–(122) can prove to be not well-posed without the demand that the required solution be of constant signs.

§5.7 Some Other Classes of Nonlinear Partial Differential Equations

5.7.1. In applications the equations of the form

$$\Delta u - G(u) = 0 , \tag{123}$$

where G is a real function given for all possible values of the required solution u, often occur.

Below we will consider some special cases of the equation (123).

The quasilinear equation

$$u_{xy} = ke^u , \quad k = \text{const.} , \tag{124}$$

is called a *Liouville's equation*.

By eliminating ke^u from the equation (124) and by using the equality

$$u_{xxy} = ku_x e^u ,$$

we obtain

$$z_{xy} - zz_y = 0 , \tag{125}$$

where

$$z = u_x . \tag{126}$$

The equation (125) is obviously equivalent to the Riccati equation

$$z_x - \frac{1}{2}z^2 = f(x) , \tag{127}$$

where $f(x)$ is any continuously differentiable function of the variable x.

The function $f(x)$ is called the *Schwarz derivative* or the differential invariant of the function $\varphi(x)$ if

$$f(x) = \frac{\varphi'''(x)}{\varphi'(x)} - \frac{3}{2}\frac{\varphi''(x)^2}{\varphi'(x)^2} .$$

the function

$$z(x,y) = \frac{\varphi'''(x)}{\varphi'(x)} - \frac{2\varphi'(x)}{\varphi(x)+\psi(y)} , \tag{128}$$

where $\psi(y)$ is an arbitrary continuously differentiable function, is the solution of the equation (127).

Replacing $z(x, y)$ of (128) into (126) we obtain by integration the solution of the equation (124)

$$u(x, y) = \log \ \varphi'(x) - 2\log[\varphi(x) + \psi(y)] + \log \psi'(y) + \log \frac{2}{k} \ ,$$

or

$$e^u = \frac{u}{k} \frac{\varphi'(x)\psi'(y)}{[\varphi(x) + \psi(y)]^2} \ . \tag{129}$$

The formula (129) has first been obtained by Liouville.

We now consider the equation

$$u_{xx} + u_{yy} = 4ke^u \ , \quad k = \text{const.} \tag{130}$$

Writing the equation (130) in the variables

$$z = x + iy \ , \quad \bar{z} = x - iy \ , \quad v(z, \bar{z}) = u\left(\frac{z + \bar{z}}{z} \ , \ \frac{z - \bar{z}}{2i}\right) \ ,$$

namely, in the form

$$v_{z\bar{z}} = ke^v \ ,$$

we conclude on the basis of (129) that the function $u(x, y)$ defined by the equality

$$e^u = e^v = \frac{2}{k} \frac{\varphi'(z)\overline{\varphi'(z)}}{[\varphi(z) + \overline{\varphi(z)}]^2}$$

is its solution for any analytic function $\varphi(z)$ of the complex variable z.

5.7.2. The equation

$$u_{xy} - \sin \ u = 0 \tag{131}$$

is known as the the *sine-Gordon equation*.

To construct some wide class of exact solutions of the equation (131) appears difficult. However, if we look for its solution in the form

$$u(x,y)=\varphi(\alpha x+\beta y)\ ,\quad \alpha=\text{const.}\neq 0\ ,\quad \beta=\text{const.}\neq 0\ ,$$

then for determining the function φ we obtain an ordinary differential equation

$$\varphi''(z)=\frac{1}{\alpha\beta}\sin\ \varphi\ , \tag{132}$$

where $z=\alpha x+\beta y$.

Multiplying (132) by $2\varphi'$ and integrating it, we shall have

$$\varphi'^2=-\frac{2}{\alpha\beta}\cos\ \varphi+\frac{2c}{\alpha\beta}\ , \tag{133}$$

where c is an arbitrary constant.

It is easy to check that the function $\varphi(z)$ determined from the equality

$$\int_0^{\varphi}\frac{dt}{\sqrt{c-\cos\ t}}=\pm\sqrt{\frac{2}{\alpha\beta}}z+c_1\ ,$$

where c_1 is an arbitrary constant, is the solution of equation (133). The function $u=\varphi(\alpha x+\beta y)$ obtained in such a way will be a real solution of the equation (131) if $c\geq 1$ and $\alpha\beta>0$.

5.7.3. Let us now consider the non-linear parabolic type equation

$$u_t=\Delta u+f(u)(\nabla u)^2\ , \tag{134}$$

where t is the time, Δ and ∇ are the Laplacian and the nabla operates in spatial variables $x_1,\dots,x_n$, respectively, and $f(u)$ is a given function defined for all the values of the required solution $u(x,t)$.

By a direct check we can easily see that by the replacement

$$v=\int_{\tau_0}^{u}\exp\left(\int_{\tau_0}^{\tau}f(z)dz)d\tau\right)\ ,\quad \tau_0=\text{const.}\ , \tag{135}$$

the equation (134) will turn into the linear heat-conduction equation

$$v_t=\Delta v\ . \tag{136}$$

Since the structural and qualitative properties of the solutions for the equation (136) are well known, the equality (135) will be able to play a fundamental role when investigating properties of solutions for the equation (134).

Generally speaking, the specification of the function $f(u)$ associated with the equality (135), which characterizes the dependence between the required functions u and v, is quite complicated. Therefore, we are naturally limited to considering some concrete examples of the equation (134).

Suppose that $f(u)$ is the sum of a Laurent series

$$f(u) = \sum_{k=-\infty}^{\infty} a_k u^k ,$$

where a_k are known numbers.

From the formula (135) we have

$$v = \int_{\tau_0}^{u} \tau^{a-1} \prod \exp\left(\frac{a_k}{k+1}\tau^{k+1}\right) d\tau , \tag{137}$$

where $\prod$ denotes an infinite product and k takes all the values of integers except $k = -1$.

In particular from the formula (137) we see that the boundary conditions of either the first boundary value problem or the Cauchy-Dirichlet problem

$$u(x,0) = u_0(x) , \quad x = (x_1, \dots, x_n) \in E_n , \tag{138}$$

in half-space $t > 0$ with the boundary E_n for the equation (134) generate the corresponding boundary conditions for the solution $v(x,t)$ and numbers a_k need to be chosen to guarantee that the problem (134)–(138) is well-posed, such that, say, the boundary condition of the Cauchy-Dirichlet problem,

$$v(x,0) = \int_{\tau_0}^{u_0(x)} \tau^{a-1} \prod \exp\left(\frac{a_k}{k+1}\tau^{k+1}\right) d\tau = v_0(x) , \tag{139}$$

for the equation (136) is not deduced from the unique solution class of the latter problem.

If the function $v_0(x)$ satisfies the condition for well-posedness of the problem (136)–(139), then by making use of the formula (137) in Chapter II we have

$$v(x,t) = \frac{1}{(2\sqrt{\pi t})^n} \int_{-\infty}^{\infty} v_0(\xi) e^{-\frac{|\xi - x|^2}{4t}} d\xi \ ,$$

where

$$|\xi - x|^2 = \sum_{i=1}^{n} (\xi_i - x_i)^2 \ , \quad d\xi = d\xi_1 \dots d\xi_n \ ,$$

which gives the solution of this problem, we obtain the required solution $u(x,t)$ of the problem (134)–(138) as a result of converting the equality (137).

When coefficients a_k for all k except for -1 and 0 are equal to zero and $a_{-1} = m$ is an integer, then the solutions u and v of equations (134) and (136) are connected with each other by the formulae

$$u^{m+1} = v \ , \quad m \neq 1 \ \text{ and } \ a_0 = 0 \tag{140$_1$}$$

$$e^{a_0 u} = v \ , \quad a_0 \neq 0 \tag{140$_2$}$$

$$u = e^v \ , \quad m = 0 \ \text{ or } \ m = -1, a_0 \ v = 0 \ . \tag{140$_3$}$$

In these cases, the boundary condition (139) has the form $v(x,0) = v_0(x) = u_0^{m+1}$ due to (140$_1$) and $v(x,0) = v_0(x) = \log u_0(x)$ due to (140$_3$) in the class of constant-signed solutions for the equation (134) (if $a_0 = 0$ when evidently we can consider it as positive without loss of generality). The unique solution class of the problem (134)–(138) can be determined easily in the case of (140$_1$). The unique solution class can obviously be determined for the case $a_k = 0, k < -1, m > 1$, as well as for the case (140$_2$), but there is already no requirement of constant signs for the solutions. The investigation becomes considerably more complicated for the equation (134)–(138) when at least one of the coefficients a_k with the index k of less than -1 and different from zero, or $a_{-1} = -1$. For example, if $a_{-1} = -1, a_k = 0$ for $k \neq -1$, then from the formula (140$_3$) it follows that the problem (134)–(138) has an infinite set of solutions in the half-space $t > 0$ of the form

$$u(x,t) = \exp \ w(x,t) \cdot \exp\left(\frac{c}{t^{\frac{n}{2}}} \exp^{-\frac{|x|^2}{4t}} \right) \ ,$$

which tend to $w(x,0)$ as $t \to 0$, where C is an arbitrary constant and $w(x,t)$ is an arbitrary solution of the Cauchy-Dirichlet problem for the equation (136) taken from the unique solution class.

§5.8 On the Smoothness Character of Solutions for Partial Differential Equations

5.8.1. In accordance with the definition, a function with continuous partial derivatives of all orders that are contained in the equation is called a regular solution of the partial differential equation if it reduces the equation into an identity.

In §5.2. of this chapter we have proved that a solution of Cauchy's problem for the Cauchy-Kovalevski equation with analytic initial data is an analytic function. From this we conclude that the regular solution of the Cauchy-Kovalevski equation with given analytic data is an analytic function provided we make use of the uniqueness of solution for this problem in the class of regular solutions. But the analogous property is not always valid in the general case of partial differential equations. In this respect there are some exceptions with elliptic type equations. In fact, let $u(x, y)$ be a regular solution of Laplace's equation, namely, a harmonic function in some domain D_0 in the plane of complex variable $z = x + iy$. We show that $u(x, y)$ is an analytic function of real variables x, y in its regularity domain. In a circle with the centre at a point $x_0 + iy_0 = z_0$ fixed arbitrarily, $D : |z - z_0| < R$ which is situated in D_0, will be the function $u(x, y)$ we represented by the Schwarz formula as

$$u(x, y) = \operatorname{Re} \frac{1}{\pi i} \int_\gamma \frac{u(t)dt}{t - z} - u(z_0) \, , \qquad \gamma : |t - z_0| = R \, , \tag{141}$$

where $u(t)$ is the boundary value of $u(z) = u(x, y)$ on the circumference γ.

Since for $|z - z_0| < R$ the expansion

$$\frac{1}{t - z} = \sum_{k=0}^{\infty} \frac{(z - z_0)^k}{(t - z_0)^{k+1}}$$

holds, we have from the formula (141)

$$u(x, y) = \operatorname{Re} \sum_{k=0}^{\infty} \beta_k (z - z_0)^k \, , \tag{142}$$

where

$$\beta_0 = -u(z_0) + \frac{1}{\pi i} \int_\gamma \frac{u(t)dt}{t - z_0} \, , \quad \beta_k = \frac{1}{\pi i} \int_\gamma \frac{u(t)dt}{(t - z_0)^{k+1}} \, , \quad k = 1, 2, \dots \, .$$

Grouping the terms on the right-hand side of the formula (142) correspondingly (this is right because of the absolute convergence of the power series in the circle D), we obtain a series with powers of non-negative integers in $(x-x_0)$ and $(y-y_0)$:

$$u(x,y)=\sum_{k,l=0}^{\infty}\gamma_{kl}(x-x_0)^k(y-y_0)^l\ , \tag{143}$$

the coefficients of which are computed by the formula

$$\gamma_{kl}=\frac{1}{k!l!}\frac{\partial^{k+l}u(x_0,y_0)}{\partial x_0^k\partial y_0^l}\ .$$

The power series (143) converges in the parallelpiped $|x-x_0|<\gamma_1$, $|y-y_0|<r_2$ where $r_1^2+r_2^2<R^2$, because the convergence radius of the series on the right-hand side of the formula (142) is not less than R at all events. Therefore near every point z_0 in its regularity domain the harmonic function $u(x,y)$ is represented in the form of a sum of an absolutely convergent power series, i.e., it is an analytic function of variables x and y.

The following more general statement is valid: if the equality (1) in Chapter I is an elliptic system of partial differential equations and the vector F depends analytically on all its arguments, then the solutions of this system are analytic functions in their regularity domain.

We have proved in §2.8 of Chapter II that the solution $u(x,t)$ of the model parabolic type equation (119) (heat-conduction equation) has derivatives of any orders with respect to all its independent variables in its regularity domain.

Suppose now that $u(x,t)$ is a regular solution of the string oscillation equation

$$u_{tt}-u_{xx}=0 \tag{144}$$

is some domain D in the plane of variables x and t.

One cannot affirm that the derivatives over the two orders of the solution $u(x,t)$ of the equation (144) are continuous near a point $(x_0,t_0)\in D$ which is in the regularity domain of $u(x,t)$. In fact, let a segment ab of the axis $t=t_0$ lie in D and

$$u(x,t_0)=\tau(x)\ ,\quad u_t(x,t_0)=\nu(x)\ . \tag{145}$$

The function $u(x,t)$ as the solution of the Cauchy problem (144)–(145) is determined by the d'Alembert formula

$$u(x,t) = \frac{1}{2}[\tau(x+t-t_0) + \tau(x-t+t_0)] + \frac{1}{2}\int_{x-t+t_0}^{x+t-t_0} \nu(\xi)d\xi \tag{146}$$

in some neighbourhood of the point (x_0, t_0).

Since the regularity of the solution $u(x,t)$ of the equation (144) means that it is continuously differentiable up to the second order inclusively, from (146) we cannot affirm the existence of higher order smoothness than two of $u(x,t)$.

5.8.2. We will says that a point x^0 in the specification domain G of a partial differential equation is an isolated singular point of its solution $u(x)$, if this solution is regular in a neighbourhood of x^0 except for the point x^0 where it may not be defined at all. The behaviour of a solution of a partial differential equation near an isolated singular point depends essentially on the type of the equation under consideration.

Let the centre α of the circle $D : |z - \alpha| < \sigma$ on the plane of complex variable $z = x + iy$ be an isolated singular point of a function $u(x,y) \equiv u(z)$, which is harmonic in D. Conjugate with $u(z)$ harmonic function $v(z)$ is given by

$$v(z) = \int_{z_0}^{z} -u_\eta d\xi + u_\xi d\eta + C \ , \quad \xi + i\eta = t \ , \tag{147}$$

where C is an arbitrary real constant, and the path of integration connecting a fixed point $z_0 \in D, z_0 \neq \alpha$, with a variable point $z \in D, z \neq \alpha$, lies in D and does not pass through α.

When the path of integration on the right-hand side of (147) goes around the point α for N times, then the function $v(z)$ will obtain an increment $2k\pi N$, where k is a real number. Evidently, the function

$$kF(z) = u(z) + iv(z) - k \ \log(z - \alpha) \tag{148}$$

is univalent and analytic at every point in D except the point $z = \alpha$, where it has a singularity, and furthermore, the expression

$$\phi(z) = (z - \alpha)e^{F(z)} \tag{149}$$

in D is an univalent and analytic function of complex variable z with an isolated singularity at the point $z = \alpha$. By virtue of (148) and (149) we have

$$u(z) = k \ \log|\phi(z)| \ . \tag{150}$$

At first, suppose that near the singular point $z = \alpha$ the function $u(z)$ is bounded. On the basis of (150) we conclude that the point $z = \alpha$ is neither a pole point nor an essential singular point for the function $\phi(z)$. Thus, for $\phi(z)$ the point $z = \alpha$ is a removable singular point and $\phi(z)$ will become be analytic in the whole D provided it extends to $z = \alpha$ as $\lim_{z \to \alpha} \phi(z) = A \neq 0$. From this, it follows that by (148) the function $u(z)$ extended to the point α as $\lim_{z \to \alpha} u(z) = k \log A$ is harmonic in the whole D, namely, $z = \alpha$ as a singular point has been removed.

Now let it be known that

$$\lim u(z) = \infty \tag{151}$$

when z tends to the point α along any path.

Under the assumption (151) the point $z = \alpha$ can be either a zero point or a pole point of certain order for $\phi(z)$ on account of (150), namely,

$$\phi(z) = (z - \alpha)^m \psi(z) , \tag{152}$$

where the function $\psi(z)$ is analytic in D and different from zero at the point $z = \alpha$, and m is a non-zero integer. It follows from (150) and (152) that near the point $z = \alpha$ the function $u(z)$ has the form

$$u(z) = k^* \log |z - \alpha| + u^*(z) ,$$

where $k^* = km, u^* = k \log |\psi(z)|$.

What remains is to observe the case that $u(z)$ is not bounded near the point $z = \alpha$ and the equality (151) does not hold. It is clear that $z = \alpha$ is now an essentially singular point for $\phi(z)$, and, hence, by the well-known *Sohotski-Weierstrass theorem*, for any real number B there exists a sequence of points $\{z_k\}, k = 1, 2, \dots$, such that

$$\lim_{z_k \to \alpha} u(z_k) = B .$$

The property just started above of harmonic functions is also valid when the number of independent variables $x_1, \dots , x_n$ is more than two. Particularly, if $\alpha = (\alpha_1, \dots , \alpha_n)$ is an isolated singular point of a harmonic function $u(x)$ and, furthermore,

$$\lim_{x \to \alpha} u(x) = \infty$$

as x tends to $\alpha = (\alpha_1, \dots, \alpha_n)$ along any path, then near the point α the equality

$$u(x) = k^*|x-\alpha|^{2-n} + u^*(x)$$

holds, where k^* is a constant and $u^*(x)$ is a harmonic function.

A solution $u(x,t)$ of the hyperbolic-type equation (144) cannot have an isolated singularity in its domain of definition. In fact, assuming that $u(x,t)$ is a regular solution of the equation (144) in the whole domain D except the point (x_0, t_0), we can affirm that the initial data $\tau(x)$ and $\nu(x)$ in the conditions (145) are continuously differentiable twice, and once respectively on the whole segment ab of the axis $t = t_0$ except the point $x = x_0$. The breakdown of smoothness for this function at $x = x_0$ implies breakdown of smoothness of $u(x,t)$, by virtue of the formula (146), not only at the point (x_0, t_0) but also at all points of the lines $x + t = x_0 + t_0$ and $x - t = x_0 - t_0$ which go through the point (x_0, t_0). These lines are characteristic lines of the equation (144). Therefore, the singularity of the solution $u(x,t)$ at the point (x_0, t_0) spreads along the characteristic lines of the equation (144) which go through this point.

By repeating the argument just stated, which is based on the formula (146), we come to the conclusion: If a solution $u(x,t)$ of the equation (144) is continuous in a neighbourhood of a point $(x_0, t_0) \in D$ and at least one of the derivatives u_x and u_t has a discontinuity, then this discontinuity will extend along the characteristics $x + t = x_0 + t_0$ and $x - t = x_0 - t_0$.

5.8.3. Let now $u(x,t)$ be a solution of the equation (144) such that, at each point (x,t) of a smooth curve σ in the domain D, it is continuable prolongable with its first order derivatives, namely,

$$\lim u(x,t) = \tau\ , \quad \lim \frac{\partial u}{\partial l} = \nu\ , \tag{153}$$

where l is a unit vector which is given on σ and not tangent to σ. If σ is a Lyapnov curve, $\tau \in \mathrm{C}^{2,0}(\sigma), \nu \in \mathrm{C}^{1,0}(\sigma)$, and, in addition, σ is not tangent to the characteristics of the equation (144) at each of its points, then a regular solution of the Cauchy problem (144)–(153) may be written in the form of a quadrature near σ. From this we conclude that a curve in the domain of definition of the solution for the equation (144) must be a characteristic curve of the equation (144) as long as the second derivatives of this solution have a discontinuity.

Not only a solutions of the wave equation, but also a solutions of a general hyperbolic-type equation, is usually called a *wave*. As in the equation (144), a wave in the domain of definition may either be "strongly" discontinuous when the solution itself has a discontinuity, or "weakly" discontinuous when its first order partial derivatives explode or break, and in the case of a linear hyperbolic equation, the discontinuity spreads along the characteristics of the considered equation and the principal reason for their occurrences is the existence of discontinuity in the initial data and their carrier. We shall say that there is a "shock" wave if a solution of hyperbolic-type equation has a discontinuity.

It is different from linear hyperbolic equations in that discontinuity in nonlinear hyperbolic equations may occur in some cases where the initial data are smooth enough and far from their carriers, as we have already shown in §5.5. of this chapter by the example of the problem (95)–(97) when $a(u) = u$ and $u_a(x) = x$.

It can be seen from the Poisson formula, which gives a solution of the Dirichlet problem for a harmonic function in a circle, that the breakdown in smoothness of the required function on the circumference will not bring about a breakdown of its smoothness inside the circle (as we have already known that a harmonic function is an analytic function of independent variables in its regularity domain).

Now let a function $u(z) = u(x, y)$ be harmonic in the whole domain D except for a curve σ in D, and $u \in \mathbf{C}^{1,0}(D)$. Because the expression $u_x - iu_y$ as a function of the complex variable $z = x + iy$ is extended analytically to the whole D through σ, the function $u(x, y)$ is harmonic in D. If σ shrinks to a point $z_0 \in D$, as we have known already, this point as a singular point for $u(x, y)$ can be erased. By this reason, the isolated points in D as boundary components cannot be carriers of continuous boundary conditions given arbitrarily. It has been proved that in the Euclidean space E_n an interior boundary component σ with a dimension of less than $n - 1$ of a bounded domain D cannot be a carrier of a given function $u(x)$ which is harmonic in the whole D except for σ under the assumption that $u(x) \in \mathbf{C}^{0,0}(D \cup S)$.